"So *you* want to become a *World-Class Leader?*"

"Patrick Hough shows us how to question our instincts and thought processes - enabling us to make better decisions – a thought provoking manual that will change the way you lead."

Alan Mc Knight
General Manager, Cara Partners

"An interesting, informative read. It explains why subjective decision-making is so dangerous, and so common. It pulls no punches, as it cuts through the jargon and reveals exactly what you need to do about it. It's an eye-opener, a must-have for anyone who wants to run an effective and efficient organisation."

John Ryan
Technical Support Director

I048111Ь

"The author asks you to suspend your views on leadership, of leadership thinking and of the core function of leaders - decision making - and the leaders decision-making process. This book challenges the status quo. It sets the reader a challenge to go on a journey and in doing so presents a wonderful opportunity to improve your decision making process and your leadership style."

JJ O'Connell
Chairman of the Plato Ireland Business Network
& Family Business Ireland

"The author has drawn on his vast management experience to put together an excellent book filled with good advice on how to maximize the potential of your organization."

Niall O Callaghan
Merck Millipore Production Manager

How to be a

World Class

Leader

USE SCIENCE TO BECOME ONE OF THE TOP

EXECUTIVES IN THE WORLD

Patrick Hough

WORLD CLASS SOLUTIONS
CORK, IRELAND

PUBLISHED BY
World Class Solutions
Cork, Ireland

Proofread and edited by James Millington
(www.jamesmillington.net)

How to be a World Class Leader / Patrick Hough. - 1st ed.
ISBN 9781983272295

Contents

1. Benefits   3

2. Mistakes   7

3. Your Role   12

4. Thinking   24

5. Fast & Slow   35

6. Personalities   44

7. Numbers   56

8. Metrics   77

9. Forecasting   87

10. Stupidity   99

11. Decisions   118

12. Control   140

13. Stories   152

14. Pricing   167

15. People   179

16. Change   202

17. Risk   215

18. Strategy   235

19. Marketing   261

20. Training   277

21. The Future   282

22. Core Tools   293

23. Execution   305

DEDICATION TO MY WONDERFUL WIFE AND FAMILY.

Can you imagine what they've had to put up with?

WARNING – Read This First

This isn't your typical business book. They follow an unspoken rule: don't antagonize the reader. This one breaks that rule—to bits.

Think of the brutal training that shapes elite special forces—the SAS, Navy SEALs. Most don't make it. This book is the intellectual equivalent, and it's designed for leaders. Or those who want to be.

You'll fall into one of three groups:

Group A – the top 10%. True leaders. They're confident, not cocky. Open to new ideas. Calm under pressure. They don't need this book—but they'll follow the rules, finish it and gain from it.

Group D – the bottom 10%. Arrogant. Defensive. Closed off. They think they know everything, and they blame others when things fall apart. This book won't help them. They won't let it.

Group B-C – the 80% in the middle. They're not hopeless, but they cling to old ideas and resist change. They feel they should be doing more, but they're unsure how. This book is for them. If they follow the rules, push through the discomfort, and do the work, they'll rise and join Group A.

Here are the four rules:
1. Start at the beginning and read to the end.
2. Skip nothing. Not a word.
3. Answer every question—after thinking hard.
4. Write all your answers down.

Break a rule, and you fail. Stop.
But a wrong answer? That's progress. A sign you're learning.

Don't turn the page unless you're ready to commit.
It won't be easy, but it *will* be worth it.

INTRODUCTION

Why aspire to become a *World Class Leader?* There are many positive reasons, but let's start with a compelling negative one: if you don't, your organization could become another statistic.

Did you know that 66% of businesses fail in the first ten years? And, in case you think those are growing pains, just check out the following list and identify the common factors:

> Woolworths, Atari, Toys R Us, Blockbuster, Kodak, Borders, Game, Clinton Cards, Jessops, Compaq, Barings Bank, Lehman Brothers, Wang Labs, WorldCom, Napster, Comet, Planet Hollywood, Ferranti, Equitable Life Assurance Society, Commodore International, Maplin, Pets.com, Zavvi, DeLorean, Pan American World Airways, Eastern Airlines, Enron.

That's a small sample of the many established companies, with considerable management talent, that ran into difficulties. Most have disappeared. You might think that the leaders of these companies were unlucky, but the truth is exactly the reverse. They had the resources and the opportunities to make their own luck. Instead, they failed and were forced to face the consequences.

This book explains the reasons behind such business catastrophes. It explains why most CEOs, senior executives and business owners are *destined* to make a decision that will lead to disaster. And it explains how *you* can be different – how you can become *World Class*.

1   BENEFITS

How would you react if you were told that almost everything you know about being a leader is wrong? *Stop. Take some time to analyse that reaction.* Are you insulted or intrigued? Will you *instantly* dismiss an idea if it conflicts with your beliefs or are you open to change? Only 10% of the people who read this book will get the *maximum* benefit from it. They're the ones who're most likely to become World Class leaders and achieve long-term sustainable progress. At the other end of the scale, another 10% will dismiss every idea and carry on exactly as before. They'll avoid everything that challenges their worldview. Which profile best describes you?

Sources

The knowledge in this book is based on the findings of leading research-ers in fields like psychology, neuroscience, behavioural economics and management science. Nobel prize winner *Daniel Kahneman* and his late partner *Amos Tversky* have uncovered some amazing insights into how the mind works and how decisions are made. *Mats Alvesson* and *Andre Spicer* have written an entire book on stupidity in the workplace. Neuroscientist *Dean Burnett* has collated research that exposes the hidden limitations of our brains, and particularly our memory. There's input from *Michael Porter* and *Richard Rumelt* on strategy. *Dylan Evans* has compiled evidence on uncertainty and risk management, and *Philip Tetlock* has developed a better way to forecast the future. There's much more. There are hundreds of researchers – too many to mention – who have contributed significantly to the knowledge in this book. Because of their work (firmly grounded in *research* and *evidence*) we know what leaders *should* do. Unfortunately, it's not what most *are* doing.

Experience

Some of the knowledge in this book comes from my own 25+ years' experience as a management consultant. I've been lucky to work with scores of CEOs in a variety of industries. I've helped billion-dollar multinationals and collaborated with employees at all levels. During that time, I've observed *the difference between management perception and reality.* I've seen managers with great intentions make decisions that work against their organization's interests. I've seen them embrace wishful thinking and dismiss logic.

Experience is useful – it's the reason I mentioned it – but it's not enough. Raw experience leads to mistakes, and the *more* experience you have, the less likely you'll identify and correct them. Are you familiar with Scott Adam's *Dilbert* comic strip? It follows the adventures of a mild-mannered engineer in a dysfunctional company. It highlights the insanity of a typical workplace and it's popular because anyone who's worked in *any* organization has come across similar situations. They can relate to the madness! Do *your* employees feel that way?

Why?

So, why did I write this book? First, because there isn't anything else like it. There are many excellent books available, many of them aimed at business leaders, but none address the foundational issues covered here. Second, I believe it can make a difference. CEOs, executives and business owners play a very important role in society. They have the power to affect the well-being and prosperity of thousands of people. They directly affect employees and their families, but also customers, suppliers and essential service providers like hospitals who depend on the taxes paid by the organization and its employees. If this book can help improve performance in an organization like yours, the benefits to society could be significant.

What's In the Book?

This book answers many key leadership questions, including:

- Why do people often make terrible decisions?
- Why do they keep repeating their mistakes?
- Why do they resist change?
- Why can't they learn?

And there's more:

- Why do they rarely complete projects on time?
- Why are they so bad at predicting the future?
- Why do they squander opportunities?

Now, as you were reading through that list, were you thinking about *other* people who possess those traits, or did you realise that "people" includes yourself?

You'll find many probing questions throughout the book, and if you think about each one carefully, you'll gain important insights about your own interpretation of the world. This book examines the underlying reasons *why* things happen and provides tools and strategies to tackle them. For example, it examines the latest research about how people *think*, and it uses that information to develop a reliable decision-making process. It also tackles fundamental questions about "leadership".

The Guide

This book will guide and support your work. It will:

- change your perception of your own role
- give you tools to become more effective and efficient
- identify tactics to improve your organization
- help you develop realistic and effective strategies

There are many potential benefits - but you'll also face a challenge. The book is easy to read, but you'll get optimum results if you let the information sink in. Schedule time to *really* think about what's being suggested and how it could affect *your* management and leadership style. Consider how each tool and tactic could be applied in *your* organization.

This book will challenge you. At times, it may even provoke you. If you find yourself getting angry, just remember: *this is helping*. You're gaining valuable feedback. It won't kill you, and it could make you stronger – so keep reading!

Who's This Book For?

This book was written specifically for you, the *individual* who makes major decisions in your organization. You may be called the chief executive officer (CEO), the managing director (MD), the site leader or the business owner. The title doesn't matter. If you're the person who's ultimately responsible for strategic, tactical and operational decisions in your part of the organization, then this book is for you. From now on, I'll just call you the CEO.

But are you a *World Class Leader*? I can't be sure. Most CEOs aren't really leaders – they just follow the pack. They've been given authority, but they don't really understand what to do with it. This book will clarify what a real leader *should* do.

Responsible

It's important to stress that *you* are responsible for how you use the information in this book and any consequent results. Only you can ensure that it's suitable for your needs, so please think carefully about each suggestion before using it. If things go wrong, *you're* the one responsible. That's *why* you're the CEO.

2 MISTAKES

In this chapter, we'll look at a list of mistakes that *average* CEOs tend to make. Do *you* make many mistakes? If you answered "no", it's likely you don't realise you're making them. That's because of the way your brain works. It's always trying to minimize the impact of unpleasant experiences. So, read through the list with an open mind and identify any that make you uncomfortable. Later in the book, you'll discover why that might be.

MISTAKE 1: Unable to Manage Risk

Some CEOs are *risk-adverse*. They ignore opportunities to increase the company's value because they're afraid of failure. Risk-adverse CEOs don't create enough value for their stakeholders. Growth is slow and when they're faced with serious competition, they don't have the resources to stay ahead. These companies often enter long periods of decline.

Other CEOs are *risk-seeking*. They're willing to gamble with their company's resources, even when the odds are stacked against them. If they succeed, they'll make impressive gains but they'll eventually fail when a high-risk venture collapses. The longer their previous history of success, the more spectacular the fall. Neither group understands how to *manage* risk, and there are usually serious consequences. Later, we'll see how you can identify the *right* level of risk for your organization and how to maximize the gains while minimizing the downside risk.

MISTAKE 2: Making Too Many Decisions

If you're making too many decisions, you're causing problems for yourself and your organization. Much of your time is taken up with trivial matters that should be handled by others. You've become a bottleneck, and this has a negative effect on your managers. They're unable to decide anything, so everything takes longer to resolve and that's causing inefficiencies throughout the organization. It also means that *you* don't have time to concentrate on critical decisions that no one else can deal with. It's hard to delegate, but it's essential if you want to make a significant contribution to the organization's long-term development, and the solution can be surprisingly simple.

MISTAKE 3: Making Wrong Decisions

When you make decisions, it's inevitable that you'll make mistakes. However, you mightn't be aware of the magnitude of this problem. It's estimated that *up to 80% of the decisions made by CEOs are sub-optimal.* "Sub-optimal" means that a decision should be better. Sometimes a decision will be *totally* wrong, and the results will be disastrous. More often, the outcome will be semi-acceptable but there were better solutions that weren't even considered.

Researchers believe that the human brain puts *serious* cognitive obstacles in the path of making the best decisions. By the end of this book, you'll be able to identify many of those obstacles and apply the most effective solutions. You'll also see how technology will soon affect how decisions are made in your organization.

MISTAKE 4: Making No Decision

Some of the biggest disasters occur because CEOs won't or can't make a decision. They continue to provide the same products or services when customers are searching for something different. Many successful companies have been wiped out because technologies have evolved. It happened when customers moved from film to digital and from DVDs to online streaming. It can happen in *any* industry at *any* time. Sometimes CEOs aren't aware that a decision *needs* to be made, so they continue to work with an obsolete model. Sometimes they're aware of an external change but they don't accept it's going to affect *them*. Many justify their inertia by arguing that moving into a new area would affect their cash cow products or services. By the time they realise the futility of that argument, it's too late to recover.

MISTAKE 5: Misunderstanding Strategy

Many CEOs talk about strategy but don't understand what it *really* means. They waste time creating plans that have no long-term value and make operational improvements that competitors easily copy and surpass.

When you eliminate the jargon, strategy is easy to understand but it's often confused with vision statements, values statements, mission statements, goals, objectives, operational excellence and tactics. Later, we'll cut through all the confusion and give you clear guidance on the best way to create a powerful strategy that will *really* work for your organization.

MISTAKE 6: Misunderstanding Leadership

Is the term "leadership" just another buzzword or does it have real importance? Many CEOs are unsure. They're confused by the many contradictory definitions. Some believe it's just getting people to follow orders, others think it's getting employees to work harder without supervision and a few reckon it's appearing in public to promote their company. The definition you use *is* important because it determines how you'll act in a variety of circumstances. If you don't have a clear definition, you may act inconsistently. Later, we'll look at the effects of those conflicting beliefs. We'll also look at a new way to define "leadership".

MISTAKE 7: Ignoring Administration

Some CEOs and senior managers become so engrossed in "top-level" decision-making that they lose sight of what's happening in the core operations and administration areas. When that happens, the result is often a *public* disaster. The apparent cause might be dishonesty, low productivity, industrial relations or public relations problems, but the fundamental reason is that the CEO has lost touch with what's happening. There's a simple way to tackle this problem. It will ensure that the lines of communication remain open and it'll avoid the senior-management group-think that's often responsible for the decline of successful companies.

MISTAKE 8: Overlooking Essential Systems

Have you let your organization grow fat and flabby like a couch potato, or have you drilled like an elite athlete, with every muscle honed? The organization's equivalent to muscles and tendons are *systems* and *procedures*. These are essential but are often overlooked because they're not obvious or glamorous. When they do their job, things run like clockwork with no drama and no excitement. Because of their low profile, however, many CEOs forget about them. They allow systems to degrade, and fail to create new ones when they're needed. The result is a slow deterioration of capabilities – like the wasting away of muscle mass. On the other hand, if you have *too* many systems and procedures, your organization will work about as well as someone with the body of a weight-lifter trying to sprint the 400 meters. Later, we'll see how to find the right balance between too much control and too little.

MISTAKE 9: Accepting Functional Stupidity

"Functional stupidity" is defined as "*the inability to use cognitive capacities in anything other than a narrow and circumspect way*". In other words, people work away in their own world and don't consider the implications of their actions on others or on the wider organization. One example is the common practice of making sure that every budget is spent before year-end. Often, it doesn't matter what it's spent on, as long as the money is used up. Functional stupidity is responsible for massive waste. It interferes with the smooth running of the business and often prevents managers and staff from reacting appropriately to outside threats. We'll outline the steps you can take to minimise the effects in *your* organization.

MISTAKE 10: Creating a Negative Culture

An organization's culture is the set of shared values and beliefs and the behaviours that result. You can think of it as a common set of habits – and it's *heavily* influenced by the CEO's actions. If the CEO is thoughtful, evidence-driven and compassionate, then those behaviours will tend to work their way through the organization. However, if he's impulsive, reactionary and vindictive, those characteristics will also permeate and impact everyone. Sometimes, a CEO will inherit a negative culture that needs to be changed. We'll look at the best way to launch and sustain a culture change project and how to overcome the inevitable challenges.

MISTAKE 11: Reacting to Urgency

When problems arise, do you immediately drop into emergency mode? Do you pull resources from other jobs and swap people around to cover the shortfall? Eventually, the problem gets solved – but then something else crops up. And this happens over and over. There's always a new problem and you never have the time or resources to solve them all. The important word in the title of this mistake is *"reacting"*. If you simply react to problems, you'll always be caught in a vicious cycle – one that's impossible to break. This book will show you how to collapse the cycle and bring things under control.

MISTAKE 12: Believing You Don't Have Problems

Finally, let's include one of the most insidious problems: "I don't have a problem – everything is fine." This is called *denial*. It's often practised by CEOs and managers because they're afraid that the alternative could lead to unpleasantness and confrontation. We'll examine this mistake carefully, because it's one of the hardest to identify and overcome.

Improvement

This book will highlight the most *fundamental* aspects of your role as CEO. If you can identify just *one* area that could be improved – and you take the necessary steps to do so – then the resulting benefits will be substantial. The book will help you do more. It will identify and optimize your latent talents and those of your staff by stripping away many of the misconceptions that are holding you back. It'll help you develop a management philosophy that's supported by *evidence*. It'll give you tools and tactics to help you achieve your goals and it'll stimulate you to develop strategies that will have a lasting impact on your organization.

3 YOUR ROLE

You'll be asked some powerful questions throughout this book. You'll get the best results if you think about each carefully and *write down* the answer. Here are the first three questions:

- Do you find it *easy* to make big decisions?
- Do employees always do exactly as you ask?
- How would you rate yourself as a CEO?

Write your answer to each question as a percentage. For example, if you find it very easy to make big decisions, you might answer "85%" to question 1. If you think you're in the top 30% of all CEOs, you'll answer 70% to question 3.

> Please write down your answers before reading on (Q3-1).

The Effective Leader

What is the mindset of the effective leader? You'll find many answers: a visionary, innovative, flexible, paranoid, adventurous, driven. All these answers have merit but there's a core mindset that trumps them all, and that's *responsible*. The expression "the buck stops here" is sometimes bandied around by CEOs when things are going well. The same people get very quiet when things go wrong. Yet that's exactly the time when great leaders stand up and take responsibility for their own actions and the actions of their staff.

Who's Responsible?

Responsibility is a term that's often glossed over, particularly in the context of the CEO's role. Have a look at the following four scenarios:

- An engineer tweaks an algorithm so that a range of cars will pass emission tests. The real emissions levels are much higher.
- A manager buys shipments of beef without checking the contents. A high proportion are horsemeat.
- A trader in a remote office loses one billion euros in unauthorized trades.
- An employee clicks on an email link and a virus spreads across the network and locks out the entire system.

Who's responsible in each case? Yes, the answer is the CEO. Now, you can see why some CEOs go quiet. They'll argue that they have hundreds of employees and they can't be responsible for what each one does. They can punish those responsible for mistakes afterwards, but they can't prevent them happening. Those CEOs see responsibility as a one-way street. They want to take the credit when things go right but avoid the blame when things go wrong. They take the word "responsibility" to mean "authority". Their *position* allows them to make decisions and give orders, but they're not prepared to accept the consequences – they simply pass the blame on to someone else.

Post Mortem

But have they got a point? After all, how could a CEO prevent the four scenarios above? The following are some questions to consider:

- What kind of culture did the CEO cultivate?
- Were questionable practices accepted?
- Were managers afraid to look too closely?
- Did they set unrealistic goals or timelines?
- Did they use undue pressure to get results?
- Was there a way for feedback to reach the top?
- Were people selected and trained properly?
- What systems were in place to prevent such incidents?

Behaviour should never be looked at in isolation. When someone does something wrong, it's likely there were contributing factors. To illustrate the point, let's examine one of those areas in more detail.

Hiring People

Many CEOs talk about people being their greatest asset, but how were those people *selected*? In many companies, people are hired based on a CV and a few interviews. Is that the best way of finding the right person? We'll discover later that it isn't.

The CEO is *responsible* for making sure there's an effective system in place to select the right people. That system should select those who are technically competent, temperamentally suited and ethically matched to the job and the organization – and it should reject everyone else. If that system is broken, it's not surprising that many of those who are hired will make mistakes or take unfair advantage of the organization. If a CEO has overlooked problems with the hiring system, then he's *directly* responsible when unsuitable people are hired, and he's responsible for the mistakes they make.

Pass the Buck

Now, let's pretend you're a CEO who doesn't believe that. You believe it's the HR manager's job to manage the hiring system. However, it was *you* who selected her to carry out that role – so if she's not capable of doing the job properly, then *you're* responsible because you picked the wrong person. OK, maybe you didn't pick her. Maybe she was already ensconced when you took over? You're *still* responsible, because you didn't work with her to improve the system or remove her when you discovered that she couldn't do the job properly.

But maybe you weren't *aware* that the system wasn't working properly? You're *still* responsible because you ignored a key system that could compromise the integrity of the organization. As you can see, the more a CEO tries to wiggle out of his responsibilities, the more evident it is that he hasn't been doing his job – and that's obvious to everyone in the organization.

The Responsible CEO

An effective CEO *understands* her responsibilities. She understands that she'll have to answer for anything that goes wrong, no matter where it happens. That means she must *work* to understand what's happening throughout the organization, she must fix anything that's problematic and she must put *systems* in place to make sure it continues to work properly. "Responsibility" isn't an abstract term, it has practical implications for the work a CEO does and the consequences if the worst happens. If a serious incident occurs, the CEO will be expected to offer her resignation. If an incident is less serious, she'll be expected to give up her bonus or take a cut in salary. She must accept that these outcomes are possible and are *justified* – otherwise, she's doesn't really believe in taking responsibility.

Responsibility isn't about punishment. It's about who has the ultimate authority to make changes. The CEO can influence *every* system in every part of the organization, so she must be satisfied they're all working properly. After all, a single employee in a remote location was responsible for the collapse of the UK's oldest merchant bank, and the modification of a small section of code slashed VW's share price by over 30%. That's the cost of overlooking an apparently insignificant area.

Leadership

We'll discuss leadership in more detail later, but you can see that "responsibility", "accountability" and "authority" are important elements to consider. Let's establish how you feel about the subject.

- Do you regard yourself as a strong leader?
- Do you expect your managers to follow your lead?
- Do you expect all your staff to do as they're asked?

> Write down the answers to these questions before reading on (Q3-2). We'll come back to your answers later.

IMPORTANT

You're going to find questions like these throughout the book. They're important because the answers will give you valuable feedback on your thinking process and your decision-making capabilities. The best way to record your answers is by keeping a notebook alongside this book. Answer the questions as soon as they appear in the text – don't wait until later. If you answer all the questions carefully, you'll gain a deep insight into your thinking and decision-making processes and may even experience some profound flashes of inspiration. That won't happen if you skip over questions, treat them lightly or fail to write down your answers. You *won't* get the same benefit if you come back later to fill them in because you'll already have seen the answers – and you won't get the opportunity to figure things out for yourself.

There are a few *defining* moments in a person's life. These are the moments when you suddenly realise that a core belief you've accepted without question is wrong. That realization can free you from deep-rooted mental shackles and allow you to make incredible changes. I hope that this book can be the trigger for several of those defining moments – and that's more likely if you *actively* engage with the questions.

Right now, grab any piece of paper and write down your answers to the previous questions. Make a note of the *question* and *page number* (or *location* on an electronic device) so you can refer to it again. If you're stuck, write the answers in the margin of the book.

Housekeeping

This is a good place to bring up a few housekeeping matters. To simplify things and reduce tedious repetition, I'll use the following conventions throughout the book:

- I'll use male and female terms interchangeably.
- I'll refer to "products" and "services", those are also interchangeable.
- I'll assume you lead an organization with four levels:
  (1) Worker, (2) Team leader, (3) Manager, (4) CEO

The principles in this book will apply no matter what your circumstances. For example, you could be a male *business owner* with 50 employees and three management levels, providing cleaning services to local families. Or you could be a female *site leader* of a multinational manufacturing division, with two thousand employees and six management levels. The same rules apply.

Must

As mentioned earlier, the purpose of this book is to get you to change your behaviour. For that reason, you'll see sentences that contain the words "*must*" or "*should*" (e.g. "you *must* treat people with respect"). Of course, you don't *have* to do anything. In this book, the word "*must*" has the following meaning: "Based on the best evidence available, you're likely to get more positive results if you do things this way." "Must" is just a little easier to write!

What Are You Worth?

Now, let's talk about *you*. How much are you worth? No, not how much you own – *how much are you worth to the organization*? It's a question all CEOs should consider. Let's do some simple calculations. We'll start with a salary of €100K/year as a baseline – we'll get to a more realistic figure later. With some simplifications, €100K/year means you're earning around €50 an hour.

However, if you were *only* worth what they're paying you, there'd be no point in employing you – you're not adding any value. *You must be worth more* – some multiple of your remuneration. So, let's go back to basics. How much is an operator worth? You'd expect to get a significant contribution from each operator – let's say they're worth double their salary. What about team leaders? Their multiplier must be greater because they've got a wider span of control. Let's say they're worth four times their salary. What about managers? Using the same principle, let's say they're worth eight times their salary. On that basis, *you should be worth at least ten times your salary.*

Justification

Many CEOs are puzzled when they first hear about this *multiplier effect,* so let's see how it can be intuitively justified:

A worker has limited opportunity to add value to the organization. He'll probably carry out a series of pre-programmed tasks and the value-add will be built into the creation of the product or service.

A team leader will have several workers reporting to her. She'll be able to minimize waste and improve quality and efficiency within her own group. Her multiplier is based, primarily, on improvements to group productivity.

A manager will be responsible for several groups. He can introduce new procedures, systems, and technologies that could improve performance across all those groups. The multiplier is based on department improvements which may include cost reductions, improvements in productivity and product or process innovations

All managers report to the *CEO,* so she has the authority to make operational, tactical and strategic decisions that will impact every person in the organization. Her decisions could impact *every* system and process. Since she has the authority to initiate or block change, she has the greatest potential to affect the bottom line.

Hourly Figures

So, to summarise, the multipliers we'll be using are:

- Operators: 2x
- Team Leaders: 4x
- Managers: 8x
- CEO: 10x

Of course, actual figures could differ depending on the industry and the size of your organization, but the principle is the same across all industries. If you disagree with the figures, feel free to calculate your own. Either way:

> **Please write down the multipliers that apply to your organization before moving on (Q3-3).**

For now, we'll work with the ones above. Let's look at the hourly figure again. We used €50/hour as our starting point. That becomes €500/hour when we use the multiplier. Now, if you're *actually* earning €250K per year (including bonuses, pension contributions, etc.) you *should be* worth €1250 an hour to the company (again, use your own figures to get a more accurate result).

Consider that figure carefully. Is every hour of your time at work worth €1250? Do you really contribute €10,000 every day? Looking at the medium term, does your presence in the organization contribute €2.5 million each year? The point of this calculation is that your time is worth more than you think – or at least, it should be. However, if you're spending time on trivial matters, you're not contributing – and that's a costly mistake.

When CEOs are independently monitored, they're often disheartened to discover how small a percentage of their time is *really* productive. So, think about what you did for the last five days. Was it worth €50,000? Should you be delegating more work? Should you make better use of your time? A meeting with five managers could easily cost €3000 per hour – are they all *really* necessary?

MROI

The hourly figure puts things in perspective, but how can you calculate how much you've *contributed* to the organization? We need to go back to first principles. Our baseline is a "safe" CEO who doesn't take *any* risks. He's in "caretaker" mode. He fixes things that need to be fixed but doesn't contribute anything more. He's just about covering his basic salary. Let's measure his performance by creating a metric called "management return on investment" (MROI). Yes, that's the same as the multiplier in the previous section – so his MROI is 1.

Of course, the performance of the company may not reflect his lack of action. The organization may be enjoying an increase in profitability every year without the CEO having done anything to cause it. It could be due to many factors: a great product portfolio, a well-known brand, weak competition, or chance. *So, an increase doesn't count unless it's due to a specific intervention.*

Proof

If a CEO wants to demonstrate that she's contributing to the organization at an appropriate level. She needs to do three things:

- Calculate the expected results with no intervention
- Describe the intervention and the expected outcome
- Record the results

History

Let's say the company's profit results look like this:
The thick line shows the company profits for the last three years.

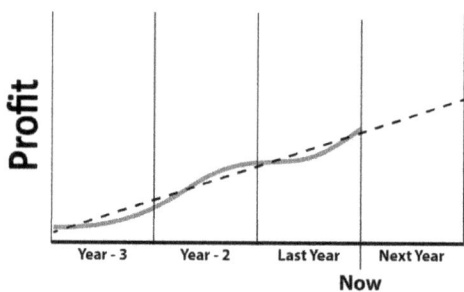

The dotted line shows the predicted results for next year using linear regression analysis. The profits are increasing in this case – but they could also be flat or decreasing. The dotted line is the baseline for next year without an intervention.

Intervention and Outcome

An intervention could take the form of a project to cut costs, to boost sales, to launch a new product or to take over a company. It could be a combination of several factors. The CEO will need to specify what's going to be done and the outcome of the intervention ahead of time. That clarification is important, as we'll see. So she might say: "We're going to introduce a company-wide programme to decrease waste. We expect an additional 10% increase in profitability by the middle of the third quarter and these savings will continue indefinitely." You'll notice that these results are on top of what was likely to happen anyway. It's an *additional* increase. Who will she make this announcement to? Ideally to everyone in the company, but only if that can be done without losing her competitive advantage. At the very least, she'll announce it to the board (the managers will already know because they've been involved in developing the plans).

Record the Results

Afterwards, the actual results will be plotted. Let's say they turn out as follows:

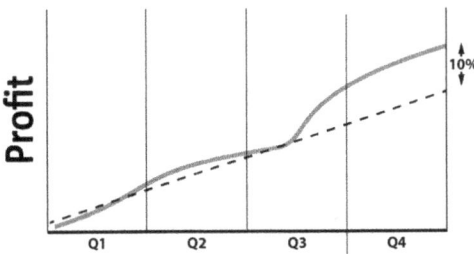

This shows that the intervention did what it was intended to do. Suppose the 10% saving – *averaged out over three years* – amounted to €3.4 million per year. Then the CEO has an MROI of 13.6 – better than our expected value of 10. Of course, many employees were involved in achieving that result, but the question is: would it have happened if the "safe" CEO was in charge? If not, then it's fair to attribute the results to her actions.

Complications

Why did we specify a three-year average in our results? Because some CEOs will try to cheat! It's relatively easy to cut costs (and boost profits) in the short term. For example, you could cut back on R&D department staff and you probably won't see the negative effects for several years. You could also cut back on the quality department, on maintenance or on brand advertising. You'll get an immediate boost in profitability but you're also eroding the viability of the business. If you take shortcuts just to improve a metric, you're creating problems for yourself and your organization.

Review

It's time for a quick review. Did you write down the answers to all the questions asked earlier? If you did you get 10 points. Those points will help you determine if you're already *a World Class Leader* or if you need to make some changes. You're ahead, so you can feel a *little* smug as you read the following paragraphs.

If you *didn't* write your answers, we need to dig deeper. I explained that if you wrote down your answers, you could expect to gain valuable insights into your decision-making process and even experience some flashes of inspiration. That could result in major benefits for your organization and yourself. You didn't do it – despite the potential benefits – so we need to figure out why.

Obviously, before you make any decision, you should examine the potential *gains* and *costs*. In this case, the gains could be significant – they could have a positive impact on your life. But what about the costs? Well, you'd have to get a pencil and notebook and then stop for a minute or so at each question to write down your answer.

There really isn't much of a cost involved, is there? On the face of it, your decision to ignore the request looks flawed. But let's list some reasons why you might have chosen that decision:

- You didn't believe the benefits are achievable
- You didn't believe it's necessary
- You don't like being told what to do
- You just couldn't be bothered

Which of these answers is closest to the truth? Take a few seconds to think about it and then:

> **Select one and write it down (Q3-4).**

Let's consider each answer. If you chose the first answer, you're being illogical. You may not believe in the benefits, but *the only way you'll know for sure is by writing down the answers.* The cost is low so there's no real downside. On the other hand, if you *don't* write down the answers, you *can't* expect to get the benefits, so you're probably wasting your time reading on. Why risk it? A similar logic applies to the second answer. If you chose either, you lose 10 points.

If you chose the third answer, you've let emotion cloud your judgement. That's one of the common reasons for Mistake 3 from the previous chapter. It's a flaw that may be interfering with your ability to make good decisions – so you lose 15 points. We'll discuss several solutions later.

If you chose the last answer, at least you're honest but you still lose 10 points because your answer calls into question your commitment to your job. People tend to have behaviour patterns that carry over from one area to another. For example, if you're a stickler for detail when buying a smartphone then you're also likely to be precise in your job. You've just taken a shortcut that could negate the benefits of reading this book, and you may be doing the same at work. You may be sabotaging your own plans because you're not prepared to put in that extra bit of effort.

Onwards

If you haven't written down the answers to the questions, please go back and do so now. If you don't have a notebook handy, you can use any piece of paper (or the margin) but it's best to transfer it to a notebook as soon as possible. It *is* important – don't start the next chapter until you've written down the answers.

Reasons

You may protest that this is a very "schoolbook-ish" approach - you *are* an adult after all. However, there are several *fundamental* reasons why it's important to write down the answers to the questions. These relate to the way you collect, process and recall information, how your mind interprets that information, and your capacity for logical thought. You'll understand all these reasons before you reach the end of the book and I promise you it's worth the effort.

4 THINKING

In this chapter, we identify how and why you make decisions. But before we start, a final check: Have you written down the answers to the previous questions? If you have, you're ready to move on.

If you haven't written down the answers – despite the reminders – then you should *stop reading now*. The reason is that you'll only get one chance to read this book for the *first* time. If you follow the instructions, you'll discover things about yourself and your job that could lead to positive change. But that feedback mechanism won't work on the second or third pass, so you'll have lost the opportunity forever.

Why *haven't* you written down your answers? We've covered several possible reasons in the last chapter, but there's one more that's more likely now – *you're afraid to admit to your mistakes*. You're refusing to commit answers to writing because you're afraid they might be wrong. Of course, that's missing the point – a World Class CEO *must* be willing to take risks and accept failures. In fact, *if a CEO can't accept that she makes mistakes, she'll never be able to do her job properly.*

So, if you're still not convinced and you're still unwilling to write down your answers, I suggest you stop now. Put the book aside and leave it. I mentioned that 10% of CEOs won't benefit from this book and – right now – you fall into that category. It's best to avoid wasting its potential.

Just because you stop doesn't mean you'll *never* use the book. You may find, after a particular incident, that you're willing to change. In that case, you'll still be able to benefit from the feedback and the insights that follow. But, for the moment, it's best to say goodbye!

If you're still reading this, I'm assuming you're on board and ready to get stuck in. So, let's begin by investigating the most important resource you have available – your mind.

Judgement

At the start of the last chapter, you were asked to rate yourself as a CEO. Now, I'd like you to answer a related question. Are you a good driver? What percentage of other drivers are you better than?

Write your answer as a percentage before reading on (Q4-1).

You may be aware of studies that show how people rate themselves when asked questions like this. For example, in one international study, student drivers were asked to assess their driving skills. 93% of a U.S. sample and 69% of a Swedish sample put themselves in the top 50%. These results are usually attributed to *illusory superiority*. This is a *cognitive bias* in which people overestimate their abilities relative to another group or the entire population.

However, there's another way to interpret these results. Did you rate yourself in the top 30%? If so, *what does that mean?* The question is somewhat ambiguous, and evidence suggests that people answer ambiguous questions by focusing on just one or two areas – usually their strengths. As a CEO, you *should* be aware of that tendency. So, when you were asked the question, you should have identified the ambiguity, broken the questions down and rated yourself individually on criteria like:

- Reaction time
- Safety
- Knowledge
- Speed
- Anticipation
- Consideration

If you'd done that, you'd have been able to identify your strengths and weaknesses and would have clarified your definition of a "good driver" – but did you? It's likely you were satisfied to answer the question as presented. That means you're willing to answer an *ambiguous* question without getting clarification. That suggests that you're also likely to accept ambiguous information from employees – and that could lead to serious misunderstandings.

Let's go back to the "good CEO" question. We need to break that down to remove some of the ambiguity. Rate yourself individually on the following:

- Forecasting ability
- Planning ability
- Decision-making speed
- Ability to motivate people
- Ability to interpret information
- Ability to prevent or avoid problems
- Ability to delegate

Write each answer as a percentage before reading on (Q4-2).

Feel free to add additional criteria We'll review these results later.

Information Glut

From the moment you get up in the morning to the minute you fall asleep at night, you're faced with a glut of information. And while your senses can detect all the information, your brain just can't handle it. That's why you've got a filter called *attention*. When you're paying attention to something, you're processing information about it. But when you're paying attention to one thing, *you're not paying attention to everything else*. Your world looks something like this – a little bubble of focus in a fuzzy background:

For example, think about your drive to work this morning. Can you remember every car you passed, every pedestrian you glanced at, every tree and bush along the way? Of course not. There's too much information to process, so most of it was filtered out. Your brain only works with things that have your attention and contain a relatively small amount of information. What attracts your attention? Things that change suddenly. If there is a loud bang behind you or if something pops out in front of you, it's going to grab your attention. The brain reacts to *change* because it's a survival instinct. That noise could be a wild animal about to attack you – your brain still contains the primitive instincts of our ancient ancestors. So, don't be surprised when you get distracted by those pop-ups on your computer!

Your attention is also attracted by things that *interest* you. For example, if you're thinking of buying an Audi R8, you'll start noticing all the R8s on the road. And you can also *force* yourself to pay attention deliberately, as you do when you make a determined effort to look for pitfalls in a legal document.

What do you do with the information you process? Most of it just goes into short-term memory and is quickly lost. A small amount is transferred to long-term memory where it's available for later use. Some information, from both short- and long-term memory is used to make decisions.

Recall

Let's test your memory. Think back to any project you were *directly* involved with that was completed around six months ago.

> **Write down five problems that arose during the project and the reasons for them. Don't read on until you've done so (Q4-3).**

Decisions

You make thousands of decisions every day. For example, when you drive to work, you don't think about the hundreds of small actions you take: changing lanes, avoiding other cars, guiding yourself to your destination. If nothing unusual happens, you'll arrive without remembering much about the journey. You've been running on autopilot. How does that autopilot work? It's a set of routines – automatic responses to stimuli – that have been built up with repetition over the years. When you were learning to drive, you had none of these routines. Every trip was agonising because you had so much to think about. As you began to repeat the same set of tasks over and over, they begin to imprint themselves and work like reflexes, allowing you to concentrate on other things. Eventually, those routines became so embedded that you didn't even realise you were using them.

Most of the decisions you make – including business decisions – are automatic. For example, when you meet a person for the first time, a set of routines kick in. You evaluate the person almost instantly, but your "first impression" is based on hundreds of previous encounters. Very little of that processing is available to the conscious mind. You may instantly decide you don't like the person, but the reason isn't obvious, you just have a *gut reaction*. But is it reasonable to trust your gut?

Gut Decisions

We all use *gut decisions*. We encounter a situation and immediately size it up and reach a decision. I like that cake – I'll eat it. I like that car – I'll buy it! I like that company – what the hell, let's buy it! Sometimes the results are as we expect – sometimes not. Research has shown that gut instinct *can* be a valid way of making decisions, but only when three conditions are met:

- The situation must be structured
- Similar situations must have occurred many times
- There must be rapid feedback

If the situation is very turbulent with a lot of things changing at the same time, then using your gut is *not* a good way to make an important decision. Likewise, if the situation rarely occurs, then gut instinct is likely to be wrong. Finally, if it takes months to learn the outcome of a decision, then gut decisions should be avoided. As you can see from that list, "gut-instinct" is useful when it's been *trained*. Our brain is an excellent pattern-recognition device and can pick up cues from the environment outside our direct consciousness. However, if the environment is too complicated, or the situation hasn't been repeated often enough, or if there's too little feedback, then it doesn't get enough reliable data to learn and can't be depended upon. Most business decisions – particularly important ones – are *not* suitable for gut decisions.

But we know from research that most managers make decisions that way. Why? Because it's easy. Let's take an analogy: you're learning to play golf. There's a spotlight illuminating the tee but everything else is pitch black. You can see the ball as you hit it but then it disappears. You can't hear anything, so you have no idea where the ball has gone. Despite the limitations, you keep on practising and get very comfortable with your swing – it becomes instinctive. Now you have an important shot coming up. What'll you do? The answer, of course, is exactly what you've always done.

It may be a terrible swing, and you may be slicing every shot, but you're comfortable with it. You might argue that this isn't a good analogy because you're *always* making decisions and getting feedback. That's true, but those are short-term decisions. They're the equivalent of putting on the green. You shouldn't be using the same swing to tee off!

There's a further complication when you *do* receive feedback. When a decision turns out positively, you'll remember it clearly as an example of your great decision-making ability. When it goes badly, you'll downplay your role in the incident or blame external factors and in a few weeks, you'll have forgotten all about it.

Brain Protection

Forgotten a failure? Surely that couldn't happen? It can – and does – because our brain tries to protect us from the emotional effects of being wrong. You might think your memory records the details of every incident you've witnessed as accurately as a video recorder, but that's simply not true. Your brain works to *minimize your suffering* and *maximize your importance*. Your memories are an integral part of that adjustment process. They've been "photoshopped" so that you appear in your rightful place – as the hero or heroine of your story.

That means that anything that doesn't fit the correct narrative is either:

- forgotten ("That never happened")
- minimized ("That wasn't important")
- attributed to someone/something else ("The economy took a dive")

If you haven't encountered this phenomenon before, you're probably shaking your head and muttering that it doesn't apply to *you*. However, research has confirmed that it happens to *all* of us. In fact, the more you deny it, the stronger you're likely to be affected by it.

Now, consider what this means: *you can't depend on your memories.* There are gaps, exaggerations and distortions. Your brain has created a series of stories that aren't completely true. It's been acting as a propaganda minister on your behalf. It's like living in a totalitarian state without realising it, except *you* are the dictator.

Review

Remember Question Q4-3? Look back at your answers and see how many problems you've attributed to *other* people or *external* factors and how many were due to your own actions or inactions. It's likely that most of the problems were *not* caused by you. Strangely, that's what most people discover!

Your Filters

As mentioned earlier, your brain isn't equipped to handle the torrent of information it's faced with every day. It's erected filters to block most of that information so only a small amount can get through. That's the *only* external information actively considered by the brain. But what happens when you look at concentrated information like newspapers, the evening news or industry reports? How do you select what gets through the filters? What's the mechanism, and how does your brain use it?

Confirmation

The brain actively searches for information that *confirms its own beliefs.* For example, if you're a Republican in the US, you'll unconsciously search for information that supports cutting government benefits and taxes. You'll ignore any information that contradicts it and you'll minimize the risk of running into that information by watching the Fox Network. If you're a Democrat, you'll watch MSNBC and search for exactly the opposite information. That's called "*confirmation bias*".

This applies to *everyone*. If you're a vegetarian, you'll read articles that confirm that meat-eaters are wrong. If you are a conservationist, you'll watch shows about renewable energy, while motor enthusiasts will watch Top Gear without a thought to ecological damage. There's a clever experiment that demonstrates this bias. Subjects are told they're being tested for a disease. They must put a strip of paper in their mouth and then dip it into a liquid in a beaker. One group are told that the strip will change colour if they *have* the disease. Another group are told that it will change colour if they *don't have* it. The strip never changes colour, so what do you think happens?

As you might expect, each member of the first group dipped the paper into the liquid only once and then walked away. Each member of the second group dipped the strip into the liquid repeatedly. You might think that's only "natural", but it's not *logical*. There's no rational reason why the test should work with one dip for the first group but require multiple dips for the second. The first group were happy – the results *confirmed* they didn't have the disease. The second group *weren't* happy, *so they searched for another opinion*. That's how our brain works *all* the time. That's why you believe "alternative facts" about many things, including your organization, your management style and your decision-making, and it's why it's difficult to change your mind. Your brain is very happy with its current assessments.

Human Bias

Let's look at several biases that make it difficult for you to make rational decisions. Just remember, this list is not exhaustive. As you consider each item, try to monitor your own reaction. Are you saying: "Yes, that could apply to me" or "Not me, I'm different". As you might guess, most people believe these biases affect others but not themselves, and therefore (ironically) are *not* different.

The Self-Serving Bias

If things work out well, *you* are solely responsible for your success (I decided..., I worked hard to...). If things work out badly, it's due to external factors (the weather was bad..., customers were slow to appreciate...).

Overconfidence Effect

We overestimate our knowledge about a subject, particularly if we have a limited knowledge to start with (remember Donald Trump's plaintive cry: "Who knew that health care could be so complicated?"). Ironically, the *less* we know, the *more confident* we are. We also overestimate our ability to predict the future. The combination of these effects leads to many projects running over budget, coming in late or failing.

Action Bias

We like to be *doing* something, even when we don't know what to do. All the activity may achieve nothing – we may even be making things worse – but at least we're *seen* to be doing something.

Sunk Cost Fallacy

You've spent €450,000 developing a new product and it's almost ready for launch. A competitor has launched a far superior product that's already gone to market. It'll only take another $65,000 to launch and market your product. What would you do? Many CEOs are swayed by the money already spent. They're reluctant to accept the losses and tend to throw good money after bad. That's an emotional reaction – not a logical decision.

Cognitive Fluency

If an idea is *easy to process and understand,* you're more likely to believe that you're already familiar with it – and you're also more likely to believe it's true. People like things that are familiar because they don't require as much mental work as those that are new. Unfortunately, there's no reason why an idea that's easy to process should be true.

Availability Bias

Are you more likely to die in a car crash or as the result of a stroke?

> **Think about it for a few seconds and write down your answer before reading on (Q4-4).**

Your brain creates a model of reality based on the information it receives from many sources including, for example, the evening news. If it's *easy to retrieve a memory*, it assumes that *the item occurs frequently*, even when that's not so. If something is repeated often enough, it becomes easier to retrieve and more likely to be accepted. This gives us a skewed view of reality. So, if you answered "car crash", you've been affected by the *availability bias*. In fact, you're about sixteen times more likely to die from a stoke than a car crash. In the EU, for example, there were around 25,000 car deaths a year in 2013/2014 while there were around 410,000 stroke deaths.

Emotional Tagging

Each piece of information isn't just stored as a factual memory. There's also an emotional element attached. This tells us whether to pay attention to the information (is it important?) and how to react to it. Think about a TV program that you'd *never* watch. Can you feel the emotional force at work, pushing your attention away to something more pleasant? That's a mild form of emotional tagging. One of the mechanisms that's involved in the imprinting of emotional content is *self-interest*. We're attracted to positive memories that suit our purposes and we're repelled from memories that don't. We get attached to people and things, and they can distort the emotional tagging of a memory.

Memories also link to each other and transfer their emotional tags but sometimes the links are misleading. Two situations can have similar characteristics, so we assume they're the same. We ignore important differences that make our assumptions invalid. For example, the last time you invested in property, the price shot up and you made a killing. Now you've taken the profits, borrowed as much as you can, and invested again to repeat that success. Unfortunately, the market is about to fall but you've ignored all the negative indicators.

What Can You Do?

Humans have many inherent flaws and the first step to tackling them is to recognise they exist. If you think you're perfect, then you'll *never* change, you're not motivated to do so. Once you recognize that you're not perfect, you can start identifying situations where your biases may be affecting your behaviours and decisions. Then you can act to correct them. It's just another irony that the best leaders are those who recognize and admit their limitations and work to overcome them.

5 FAST & SLOW

In the last chapter, we had a look at the some of the biases that affect human decision-making. Now we're going to look at the nature of the thinking process itself. We normally consider ourselves as a single entity – anything else really doesn't make sense, does it? However, if you're a single entity, why do you often find yourself in conflict? You decide to exercise, but you can't get yourself to the gym. You want to lose weight, but you can't give up your favourite food. You want to watch less television and read more, but you still find yourself plonked in front of the TV. Obviously, there's a lot going on in our brain – it's not just "me" in there. It's not as simple as it first appears.

Systems 1 and 2 (Fast and Slow)

In his bestseller *Thinking Fast and Slow,* Nobel Prize winner Daniel Kahneman describes two "systems" in the brain that explains many of the features of how we think. The first he called *System 1* – this is the *fast* system, it's responsible for instant decisions. The second system – the *slow* one – he called *System 2.* You can always remember which is which because the fast one always comes first! It's easy to demonstrate the two systems in action. Do the following in your head:

4 + 5 = ?

That was the fast system in action. You've already learned the addition tables, so your System 1 could react immediately with the answer. There was no "thinking" involved. Now do the following calculation in your head:

74 x 43 = ?

This time, you experienced the slow system at work. There's considerable effort involved because you must retain a partial answer in your memory while you multiply the remaining numbers together. As you were doing so, your pupils dilated and your brain burned more energy. When you're driving to work, you're using System 1 most of the time. You don't have to "think" about your driving or your route. Afterwards, you may not even remember anything about the drive. On the other hand, if you're driving in a crowded foreign city for the first time and you're not sure where you're going, the experience is very different. Now you're using System 2 to stay on track and you feel exhausted afterwards. So, System 1 is nice and pleasant – you get answers without having to "think" about them, and there's no effort involved. It's responsible for intuition and gut decisions. System 2 is a lot more uncomfortable, it takes effort and focus and it burns a lot of energy. System 2 is responsible for logical decisions.

Fictions

Before going any further, you should know that System 1 and System 2 are both *fictions*. They're useful metaphors that describe something far too complex to discuss easily. They help us think about how humans respond and react without getting bogged down in detail. In case that concerns you, don't forget that many of the things we routinely accept are also fictions. For example, look at the following news items:

- White House denies allegations
- Google clamps down on fake news outlets
- UK votes to leave EU

Obviously, the White House is a building, so it can't deny allegations. We might assume that the report refers to an announcement by the president, but if he announced it himself, they'd probably have reported the fact. It may have been the press secretary who made the announcement, but it's unlikely that she wrote the content herself – other people were involved. Was the president even aware of it? We can't be sure.

What about the second item? Well, Google is just a legal entity. It can't say or do anything either. Someone working in Google probably said something to a reporter, but was that person authorized to make the announcement? Were they speaking at a policy level, an operational level or a technical level?

The third item is even more confusing. The UK is a territory that includes land, buildings and so on. It can't vote. Some of the people living in the UK *can* vote but many others either can't or don't. The UK population was 65.6 million in 2016. Only 33.6 million people voted in the Brexit referendum of which 17.7 million voted to leave and 16.1 million to remain. The difference between the leave and remain vote was 1.27 million – about 2% of the total population. And, of course, despite that vote, the UK is not going to get up and move away!

We have no trouble accepting fictions like those every day. We attribute human characteristics to inanimate objects like buildings, companies and countries. There can be problems doing so – we'll talk about those later. But for the moment, we're going to suspend belief and apply the same approach to the workings of our brain.

Links

Think of the word "soap". What comes to mind? You might have imagined washing your hands with a white bar. You might have "seen" the suds, felt the smooth texture, smelled the fragrance. You might have heard water trickling over your hands, washing the suds away. There was no effort involved – one sensation linked to another effortlessly. Now, try another example. Think of the word "classroom". Again, you'll experience a series of memories. Perhaps you "see" a blackboard with chalk marks or a whiteboard with markers. You can picture the desks, the teachers and the pupils.

Those examples illustrate how your System 1 works. It uses *associative memory*. One idea is linked to the next, and the next – and so on. For example, if someone says "credit", the links in your brain might light up like this:

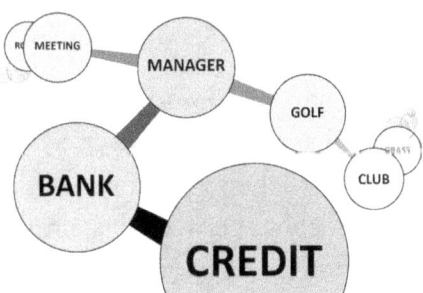

Those links activate almost instantaneously and without any effort. The nodes aren't just words; each includes an entire cluster of impressions. "Bank" might include images of marble floors, polished brass-work, smart-dressed tellers, echoing footsteps and the smell of polish. What about emotion? Perhaps apprehension or fear from the time your younger self went looking for a personal overdraft?

Emotion is tightly intertwined with our memories. If you are asked to choose a paint colour for the office, your mind is already linking the idea to a vast range of memories, most of which you're not even consciously aware of. Perhaps your father painted your room a pastel green colour when you were eight years old. You didn't get to choose the colour (you wanted red) and you sulked for days before accepting the inevitable. There could still be a link between that colour and your emotional response. So now, when asked to choose a colour, you instantly reject pastel green. Why? "I just don't like it." That's a gut decision – there's no obvious logic. You can't explain it and you can't justify it. That's System 1 at work.

Action

System 1 is always on, scanning the internal and external environments, looking for any surprises. If anything unusual happens, it's primed to react instantly. If it's been trained to act in a specific way, it'll choose that action. For example, if you're driving and a child runs out onto the road, you'll slam on your brakes "without thinking". You'll react just as quickly for other stimuli – for example, when someone disagrees with you. We sometimes talk about a person's "hot buttons" – the stimuli that will provoke a predictable reaction every time they occur. These are System 1 reactions and you may not even be aware you're having them.

Logic

System 2 is in charge of logic. It can work things out. It can analyse data and carry out difficult calculations. However, System 2 is lazy. It'll do as little work as possible. Whenever possible, it'll hand off the work to System 1. The division of labour between System 1 and System 2 works well most of the time. System 1 recognizes a situation and invokes an automatic response. There's no need to disturb System 2 because the same set of actions can be played over and over.

You have *social scripts* for occasions when you interact with people. You have a script for meeting people, for saying goodbye and for other social occasions. When you go to a restaurant, you know what to expect before you enter. You announce your presence, wait to be seated, look at the menu, and so on. Every step is predictable. You're at ease because you know what to expect and can react automatically. Occasionally a script will lead to absurd behaviour. For example, two people may pass each other on the street: The first asks, "How are you?", and the second answers, "How are *you?*" Both walk on, satisfied with the encounter even though it didn't make the slightest bit of sense.

Patterns

System 1 is good at finding *patterns*. In a restaurant, it can identify exactly what to do – and that isn't as easy or natural as it appears. After all, every restaurant is different. There are different staff, different layouts and different menus. There are hundreds of thousands of combinations. But your System 1 has identified common patterns in all those combinations.

System 1 also offers suggestions to System 2: "This is a good restaurant." "The waiters are friendly." "We should come again." These are reactions – not the results of a formal evaluation. The waiters seem friendly because they smile and chat in a way that reminds you of your friends.

Errors of Judgement

There's an obvious problem with System 2 depending on System 1 for information. System 1 is *not* logical. It reacts based on its training and previous experiences, and those stretch all the way back to childhood. It's particularly interested in its own short-term survival and its own pleasure. It searches for acknowledgement that it's been a good boy or girl. That's why it's hard to force it to concentrate on serious things like long-term health care or sound financial investments. The short-term is much more enticing. System 2 is being continually fed information from System 1. That's why much of what it works with is superficial and biased. It takes hard work to search for information that contradicts your existing views, sift through them carefully and come to a measured conclusion. It's much easier to work with what you've got, however incomplete and lopsided.

Justification

System 2 has one job that it's usually rather good at. Remember your response to the pastel green paint earlier? Your System 1 rejected it without being able to give a reason. It just didn't like it. However, if you're *pushed* for an answer, your System 2 will step in and justify it. "It clashes with the colours in the corridor." "It makes the room too small, and we need a bright colour to expand it." Sometimes the reason doesn't make any sense but sounds like it does. So, System 1 often *decides* but gets System 2 to *justify* it:

- *System 1:* I'll look fantastic in that big car and everyone will see me and be jealous.
- *System 2:* It's got low emissions, the highest safety rating, and the best MPG for a car of its class.

And:

- *System 1:* When I take over that company, everyone will be amazed. They won't believe anyone could be so clever. They'll write about it in all the papers. I'll be famous – their CEO will be crushed.
- *System 2:* The takeover will provide us with complementary technologies that will expand our customer base. The synergy from our combined R&D groups will provide a stream of innovative products that will continue into the foreseeable future.

The person involved often *believes* the System 2 version of events. They're fooled by System 1 into thinking they've made the decision using System 2!

Mental Energy

One of the most important functions of System 2 is self-control. It's responsible for making sure you stick to your diet, exercise regularly and make logical, data-based decisions. Unfortunately, as we've seen, System 2 is lazy. It *can* make a deliberate effort and keep our desires in check, and it *can* concentrate and make logical decisions, but it can also run out of steam very quickly. The problem is that each day, we only get a limited pool of mental energy. System 2 uses a *lot* of that energy when it's doing its job and the harder it works, the faster the pool is used up. Once the pool is depleted, it goes back to sleep and lets System 1 take over. That's why plans and resolutions are suddenly dropped and we take the easy way out – sometimes after months of hard work. There's an obvious lesson from that description. *You should always carry out the most important tasks at the start of the day* and leave the less important ones until later. Unfortunately, many CEOs do the reverse. They work on emails in the morning and have little time or energy for important decisions later in the day.

Distractions

Here's a tip: if you've ever been asked to think of a number while reading a short paragraph of text, you should be suspicious. The reason is that you'll be less critical of what you read and you're more likely to accept the information. That's because System 2 is responsible for critical thought and is tied up with the memory task. Of course, you can get distracted in many other ways. If you're driving and speaking on the phone at the same time, your response time will be several seconds slower if an emergency arises. If you're concerned about something in your personal life, it's also likely to negatively impact your business decisions.

Data Collection

When System 1 collects data, it uses its training and experience to decide what to concentrate on and what to accept. It reinforces existing biases. If you want to achieve more, System 2 must be involved in the process. However, System 2 won't get involved unless you make a determined effort. You must overcome its reluctance and "force" it to get involved. One way to do that is to provide a structure for the data collection and decision-making process. You'll learn about that later.

Patterns

How good are you at *interpreting* patterns? Have a look at the following description and then answer the question:

John is thirty-two years old and lives in London. He's very bright but can be arrogant at times. He has an advanced degree in mathematics and he drives a Porsche. Now, *which of the following statements is most likely*:

- He's a social worker.
- He works in a bank.
- He's a member of the Green party and plays cricket.
- He's a traditional musician and collects jazz records.
- He works in a bank and trades stocks.

Write down the answer before reading on (Q5-1).

If you haven't seen similar questions before, it's likely that you have a high degree of confidence in your answer. That's because your brain likes stories. It finds abstract thought difficult but it likes to think about personalities with habits and abilities. It also likes to find causes. It attributes good events to heroes and heroines and bad events to villains and villainesses.

So, did you choose the bank worker who trades stocks? That's understandable, because the profile certainly fits our mental picture of that type of individual. However, it's wrong. The right answer is "bank worker". Why? Because there are more "bank workers" than "bank workers who trade stocks" and the term "bank worker" includes *all* those who trade stocks – so it *must* be the most likely answer.

Now, stop and think about that answer for a minute (yes – 60 seconds). *Did you just dismiss the question as a piece of trivia?* If so, it confirms what we mentioned earlier. Your brain doesn't want to label this experience as a failure, so it found a way to downplay its importance and relevance.

In fact, it's a *very* important result because it demonstrates many of the characteristics of your System 1 and 2. It confirms that your System 2 is lazy. The correct answer should have been obvious, with a "little thought", but your System 2 probably didn't get involved. Why? Because your System 1 jumped to conclusions. It matched the description with the possible answers and *stopped once it had a good story*. It used a stereotype to make its decision.

The Good News

Nobody likes to hear that their decision-making process is faulty, or their logical faculties are lazy, but it's a *good* thing if you're willing to do something about it. If you could improve the outcomes of your decision process by 10%, 20% or even 30%, wouldn't that be worth doing?

Something New?

Now, are you ready to move on to something new? If so, that's a big mistake. *You've just been given an opportunity to learn.* Unless you take the time to dwell on it, and work out what it means, you'll forget, and the benefits will be lost. Write the mistake in your notebook (e.g. "*Failed to use System 2 because I accepted a stereotype*"). Now identify *why* you made it and write that down. Next, write down your *feelings* as you were making it. Finally, write down the *implications* if you often, unknowingly, make similar mistakes. Spend at least 5 minutes thinking about what happened. Do this every time you realise you've made a mistake in this book.

Record your answers before reading on (Q5-2).

Did you write down the second mistake? The first was accepting a stereotype. The second was being too eager to move on and not spending the time required to process a learning experience. If you didn't write *that* down you've also committed a third mistake: failing to *recognize* an important learning experience!

FURTHER READING

Daniel Kahneman, *Thinking Fast and Slow*, Penguin Books.

6 PERSONALITIES

People are watching you. Why? Because you have the power to affect their well-being. If you make good decisions, the organization will succeed and they'll succeed with it, but the reverse is also true. People also want to understand what it takes to get on in the organization. Do they need to be a "yes" person? Should they come up with new ideas? Should they challenge the status quo or is it better to avoid risk and shoot down anything that threatens it? If they want to avoid being sidelined, what's the best tactic? Do they need to tread carefully? Is it dangerous to be the bearer of bad news?

Let's take an example of how the process works. It's your weekly meeting and the managers are giving an update. Mary White, your sales manager, has just informed you that the expected increase in sales is not going to happen. That means you'll fall short on your quarterly targets and the board members aren't going to be pleased. What you do now will send a message. Suppose you lose your temper, pound the table and start shouting at Mary. What does that do?

- It tells people that you don't like bad news.
- It shows that you react *emotionally*.
- It shows that you don't *respect* people if they fail.

Of course, the news leaks out after the meeting and Mary's reputation is damaged. But Mary and her friends have lost respect for *you*. She feels she didn't deserve the public dressing down and her friends are telling people you're unreasonable (the shortfall wasn't her fault – it was due to external factors). You've just made some new enemies.

When people hear about the incident, they grow more cautious. Now, they're less likely to give you bad news. The organization's culture has become more negative, so people are less likely to take any chances. Innovation is less likely.

Let's replay the incident again. This time you *don't* lose your temper. In fact, you don't say *anything* – you just nod and carry on. What does that do?

- It tells people that failure is not a big deal.
- It tells them you're not too worried about sales figures.
- It may suggest that you don't pay too much attention.

The news leaks again but this time, it's your reputation that suffers. Some feel that you should have tackled Mary about her dismal performance. Others relax a bit. After all, if Mary can get away with it, so can they. In this case, your lack of a reaction could result in lost productivity.

You must *calibrate* your behaviour and avoid being provoked into doing anything that could be misunderstood. You should never lose your temper, because the negative effects are likely to be significant. It shows that you don't respect the victim – particularly if you do it in public. People will sympathise with them and your reputation will suffer. The cumulative effect of many incidents will inform people what's important to you and what will happen if you don't get what you want. *They'll* decide whether you respect people, whether you're supportive or whether you only care about yourself.

External Factors

Of course, CEOs are not always *consistent*. Sometimes, their reaction is influenced, not just by what they're hearing, but also by how they happen to be feeling at the time. If they've just had a fight with their spouse, they'll act aggressively. If they've just had a great game of golf, they'll be positively helpful. That uncertainty confuses people and increases their stress levels.

The Solution

Most CEOs react *spontaneously* to information. They wait for something to happen and then react. At that point, they may be in the grip of emotion and that's the worst time to decide. The solution is *to work out beforehand how to react under different circumstances.* How *should* you respond when you encounter the following?

- Sales targets are going to be missed.
- A critical project is falling behind schedule.
- Someone was injured in the operations area.
- Your best manager tells you he's leaving.
- Someone arrives later for a meeting.
- Someone hasn't completed their tasks for the meeting.

Write down the *best* response for each encounter. Then add other possible circumstances you could come across in your organization and the best reaction to each.

Behaviour

So, you don't get angry and you're careful about what you say – but things are *still* not working as they should. For example, you're always telling your managers that quality is important. You've asked them to spread the message to their staff, but you're still seeing significant problems. Why *is* that? It's probably because of an incident six months ago when you told the quality engineers to release a rejected batch so you could meet your year-end shipment figures.

You keep hearing about people not following procedures. You want the team leaders to double down on enforcement but last week, when you were showing visitors around, you removed product from the line with your bare hands, even though the procedure strictly forbids it.

One of the greatest dangers any CEO faces is *arrogance*: "The rules apply to everyone, except me." If you don't follow the rules, people will notice, and they'll follow your lead. It doesn't matter what you say. What you *do* demonstrates what's important. If you take shortcuts, so will your staff.

Leakage

If you play poker, you know about "tells". These are small physical indications that your cards are better or worse than you're pretending. It might be something as obvious as scratching your ear or as subtle as a slight intake of breath or narrowing of the eyes. Our faces are natural broadcasters of emotions and people have developed the skill to read those emotions without even realising it. Even if we try to suppress those visible indications, they may still be clear to anyone who is paying attention. And that's particularly important in an organizational setting where people sit opposite each other at meetings.

When a person is talking at a meeting, the audience will do two things. First, they'll look at the speaker and then they'll look at you to gauge your reaction. It'll take considerable effort to remain neutral if you have a strong reaction to what's being said. If you have any biases or prejudices, they're likely to be exposed, not only by your face but also by the words you choose.

Suppose you believe that people over fifty are too old to do a good job, and they're just waiting to retire. When you're talking about your older employees you may find yourself using words and phrases like: "exhausted", "worn-out", "the old days" and "tired". Your brain communicates your feelings, despite your best efforts to hide them. This is called "*leakage*" and the best way to combat it is to identify any biases you might have and confront them. Use knowledge to eliminate them.

Of course, that's not easy, because many biases come with *emotional tagging* and that'll make the exercise uncomfortable. However, it's worth persisting, because biases are illogical. You just need to prove that for yourself.

Start by questioning your belief. *Why* do you believe older workers aren't productive? Well, remember Mike? He was lazy. And Stephen didn't understand computers. You can readily remember four or five individuals who were over fifty and had problems doing their job. That's your "proof". Now, do you know anyone under fifty who's had similar problems? (Ron, Judith, Harry, Kristy.) And do you know anyone over fifty who did a great job? (Malcolm, Trina, Carl, Mark.) There are counter-examples if you care to look and that means your conclusions are flawed. *Prejudices are usually caused by using a limited data set to generalise about the entire population.*

Your Prejudices

What are *your* prejudices? Before you can tackle them, you need to be aware of them. So, carry out the following exercise and write down your answers. Do you believe that a person from the following categories is less mentally capable, less honest or less worthy than you?

- Another race
- Another colour
- Another ethnicity
- Another race
- Another nationality
- Another region
- Another social class
- Another religion
- Another educational level
- Another age group
- Another profession
- The other sex

Take some time to explore your feelings. Write down any reservations you have with people in each of these categories. Be specific. For example, if you have a reservation about people of another nationality, specify the nationality and the specific reasons. Also, write down your evidence for feeling that way.

Don't read on until you've completed this exercise (Q6-1).

Did you find that exercise difficult? If you rushed through it, go back and do it again. It's important to expose your prejudices and see them for what they are. Don't let them influence you subconsciously. Sometimes prejudice can be hidden in plain view. For example, some US citizens boast that America is the greatest country in the world. Of course, that implies that all other countries are inferior. How many Americans really believe that? Those who do will look for evidence that it's true and avoid evidence that it isn't. They're happy to let System 1 drive their thinking.

System 1 can link ideas together with little evidence. For example, if you met just one Scandinavian, and he stole your wallet, you might believe – even subconsciously – that *all* Scandinavians shouldn't be trusted. If, instead, he helped you when your car broke down, you'd have a different impression. Every sizable group has good and bad individuals. When you interact with a small number of them, you may be lucky or unlucky. However, *if you judge the entire group by the actions of a few, then you're prejudiced*. We'll be looking at a tool to help you tackle that later in the book.

Your Image

You're constantly judging other people and they're judging you – but since you've become CEO, you're more visible than ever. So, how are you currently viewed by your staff? Do they think you're:

- Strong?
- Warm and approachable?
- Flexible?
- Inclusive?
- Friendly?
- Decisive?

Write down the answer before reading on (Q6-2).

It's likely that you believe that most of your employees have a favourable impression of you. Would it come as a shock to learn that they have a *completely different* opinion? Perhaps they believe you're:

- Cold and distant
- Inflexible
- A bully
- Bureaucratic
- Weak

Do you think the CEOs who're seen as bullies *know* what people think of them? Do you think that CEOs who're regarded as weak realise it? In most cases, the answer is "no", because they interpret their own actions differently – and that could be happening to you. You might think a particular action is "strong", but employees think you're insensitive. You might believe you're being friendly and inclusive, but they think you're being weak and indecisive. The same set of actions can be interpreted as friendly or weak depending on the framing.

Changing Your Image

It's hard to change someone's impression once it's been set. You'll need to demonstrate a consistent set of behaviours over a long time to get people to alter their impression – and it could easily flip back with just one lapse. The most critical time is when something has gone wrong and a problem has arisen – that's when people will watch you closely. How do you handle it? If you get angry, the image you've tried to create of a calm and friendly leader is gone. If you take shortcuts – like ignoring procedures or quality rules – then your credibility is lost. And if you do anything that could be interpreted as dishonest, like taking unfair advantage of a customer, then your reputation is tarnished forever.

It's easy to remain calm and follow the rules when things are going well. However, if you get caught up in a crisis, you might find yourself ignoring the long-term view and concentrating on the immediate problem. This is called *perceptual narrowing*. It's a psychological condition that occurs when a person focuses so intensely on one small thing that they ignore everything else without realising it. It's been blamed for many disasters, including several plane crashes where pilots have become so distracted by warning indicators that they fail to react to the real danger.

Personality

We all have an inherent personality which has been developed over the years by numerous interactions with people and situations. "Personality" is defined as "an individual's relatively consistent behaviours, inclinations and preferences across different contexts". In other words, we automatically tend to react in the same way whenever we find ourselves in a specific situation. That's a good description of System 1 in action.

Does everyone have a unique personality? The answer is yes, because everyone is different. Despite this, psychologists have examined many individual personalities and identified certain trends. They believe it's possible to classify people into categories. There are various tests you can take to find out which category you fit into. These include the Myers-Briggs Type Indicator (MBTI), the 16PF Questionnaire (16PF) and the DISC assessment. The categories in these tests have many similarities and common features. We'll consider a similar system that looks at decision-making styles in a business environment.

Decision-making Styles

Robert Miller and Gary Williams created a categorization system based on a study of 1,700 executives over a two-year period. Their system identifies five types of people based on their decision-making styles (they argue that this is not the same as a person's personality in the broader sense). These are:

- Charismatics
- Thinkers
- Sceptics
- Followers
- Controllers

While there's no absolute right or wrong style, there's evidence to suggest that some styles are more likely to lead to bad decision-making. It's useful to know which category you fit into and the inherent strengths and weakness of each. Let's have a look:

Charismatics

Charismatics are always after the next idea, particularly big, bold ones. When they find one, they get very enthusiastic and want to jump in. They don't want to get involved in the fine detail themselves, but they'll get someone they trust to work through it before they commit. They are very enthusiastic and eager to accept responsibility for making big decisions. If they make a mistake, they're willing to take the blame.

These people use System 1 most of the time because they find System 2 cumbersome and annoying. They prefer to get other people to work through the logic.

Thinkers

Thinkers are intelligent, logical and academic. They are readers. They're interested in detail and won't make any decisions until they've worked through all the pros and cons and examined the small print. They're driven by data and information and will change their mind when presented with enough evidence. They don't usually let emotion get in the way and tend to make rational decisions.

This group uses System 2 more than any other group. System 1 is still the primary decision-maker, but they've trained it to operate in useful ways. They follow comfortable routines: collecting data, analysing it and reaching a conclusion. However, they can still be led astray by System 1, particularly when dealing with novel situations.

Sceptics

Sceptics are suspicious and will question everything, particularly if the new information doesn't line up with their existing view of the world. They are outspoken, aggressive and demanding and will frequently step on people's toes to get what they want. They're also determined and visionary and have no difficulty making big decisions. They can be aloof and emotionally detached.

These people have used System 1 to build a model of reality where their own importance and competence are assured. They use System 2 to justify and support that reality. That's why it can take a lot of convincing to change their views.

Followers

Followers like to maintain the status quo. They don't like change and will resist it whenever possible. They are responsible and serious and will work hard to maintain their organization. They work well with people and can get them to do as they want. They value feedback and will try to achieve a consensus whenever possible.

These people have a strong System 1 with highly developed social scripts. They want to fit in and don't like change because it could upset the social order. They don't need to use System 2 because they can depend on other people to do the in-depth thinking.

Controllers

Controllers are driven by fear. They're always looking for threats to themselves and their organization. They're almost totally self-reliant (because they don't really trust others) and will make major decisions without input from anyone else. They are unshakable in their beliefs.

They are also *perfectionists*. They'll get involved in the smallest detail and often frustrate those who work for them. However, while they've no problem making decisions, they'll avoid taking the blame when things go wrong. *They* are never wrong and will blame others for their mistakes.

This group's System 1 is tightly integrated with scripts linked to fear, particularly the fear of failure. This can push them to go further, to gain people's respect and admiration. However, these scripts will also prevent them from trusting other people. They can use System 2, but often it's to criticise others, justify their own actions and direct blame away.

What About You?

Do you see yourself in any of those descriptions? It's worth understanding your profile so that you can identify your weaknesses and take steps to address them. Of course, there's a problem. Your brain may want to pick the "right" profile because it's trying to hide the unpalatable truth from you. If there's any doubt, monitor yourself to see if your behaviour matches one of the "less pleasant" profiles.

Reporting to a Personality

Another way to consider each of these personalities is from the viewpoint of someone who works for them. Let's see how that goes:

If you work for a *Charismatic*, you're in luck because she's a friendly and reasonable boss. She'll look out for your welfare and give you credit when you succeed. However, you probably get frustrated with her on a regular basis. The reason? She keeps changing her mind. She might have asked you to carry out a project. Everything is going well, but then, before you can complete it, she suddenly decides there's something more important and pulls you off. This can happen again and again.

If you work for a *Thinker*, you'll find him easy and fair to deal with. However, he's very precise and will want to see data to back up any claims you make. He won't make decisions quickly and you can annoy him by going too fast without having all the information. He doesn't like sloppy work, so you need to prepare carefully for your meetings.

You'll have more difficulty if you work for a *Sceptic*. She likes things done her way and can be rude if that isn't happening. You'll have difficulty getting a fair hearing for new ideas because she won't believe what you're saying, particularly if it conflicts with *her* views. Don't expect to get credit for them either.

Followers are a joy to work for. If one is your boss, he'll be warm and friendly and will look after your welfare. However, if you want to make changes, he can be frustrating, because he'll find ways to slow your progress to a crawl. He won't like to change and will only do so reluctantly.

You'll find it more difficult working for a *Controller*. She'll try to micromanage everything you do. Everything must be done *her* way. She will make all the decisions and she'll blame you if anything goes wrong. She'll listen to your ideas but, if they're any good, she'll take the credit when they succeed.

Personality Change

Can you change your personality and your decision-making style? It *may* be possible, but only with a lot of effort. Habits are hard to change. When you're naturally suspicious or afraid of change, you'll regard your traits as "natural" and reject any alternative explanations. And since it's unlikely that anyone will challenge you within your organization, it's easy to believe that you're perfect!

Culture

We've seen how employees watch what you do, and we've seen how your System 1 reveals itself in your personality and habits. Researchers have also confirmed something very interesting: *your personality can directly affect the culture of your organization – and its profitability.*

That's an important finding because it means you'll have to change how *you* behave if you want to change your organization's culture. That's not easy, but we'll be looking at several steps you can take to make it more achievable.

FURTHER READING

Miller, Robert B., Williams, Garry A., *The 5 Paths to Persuasion: The Art of Selling Your Message*, Warner Business Books.

7   NUMBERS

CEOs often make bad decisions because they don't use *numbers* properly. In this chapter, you'll learn how to identify and avoid many common mistakes.

Graphs

Graphs are used to identify patterns in data that wouldn't be easy to spot just by looking through columns of numbers. They look simple enough, but they can also mislead and must always be treated with caution. Let's look at an example. What would you do if you were faced with this sales graph?

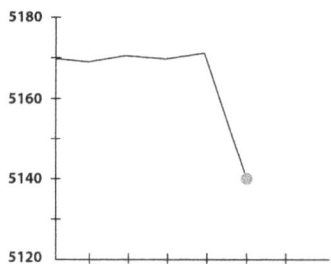

Write down your thoughts before reading on (Q7-1).

It doesn't look good, does it? It appears there's a serious problem. However, you'll notice that the Y-axis starts at 5120 rather than zero. When zero is used (next page), the drop in sales is hardly noticeable. If you want to appreciate the true magnitude of a change, you need to use a *zero baseline*.

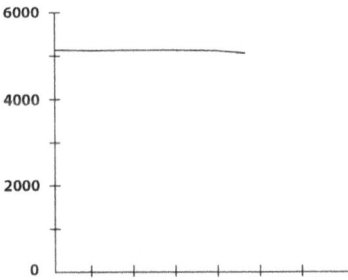

You *can* use a non-zero baseline if you want to magnify small changes, but that can often be misleading and counterproductive. *Most small changes are due to random variation* and there's a danger you'll waste time trying to fix fluctuations that can't be eliminated – you'll just make things worse by intervening. Occasionally you *will* need to improve a process because it's not operating properly but it's far more common to see results go up or down without any significant cause. This is called *natural process variation.* You can use a technique called "Statistical Process Control" (SPC) to determine if a process really needs to be fixed.

Misleading

Some people deliberately use graphs to give a misleading impression. If you glanced at the graph below, you might think that Ann has twice as many sales as Bart. But look at the scale and you'll detect a different story:

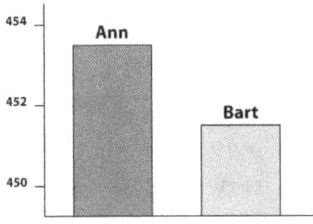

Graphs can be biased in many ways. If someone shows you a graph that has no values on the axis, it's possible they're trying to mislead you – insist on seeing the figures. People also mix different variables on the same graph without labelling them properly – always check that you're comparing like with like. Have a look at the following example and see if you can identify any issues.

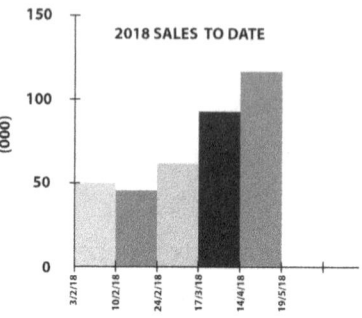

Write down your answer before reading on (Q7-2).

Did you spot the problem? This graph was designed to mislead. The time intervals are *not* the same; they increase as we move across the graph. In this case, all the information is available to detect the problem, but most people won't go to the trouble of doing so – and that's what the author is counting on.

Many graphs shown on TV and in magazines and newspapers suffer from one or more of the problems described above. It's good practice to spend a little time trying to spot them!

The Kandanesian Problem

How confident are you when you've made a decision? The following example will test your logical skills *and* your confidence in your own decision-making process.

You're interested in setting up a business in Kandanesia but the government controls all business licences and has a unique method of allocating them. You pay the licence fee up front and then you get to choose between three boxes. One contains a licence to operate a business in the capital – that's the one you want. The other two contain licences for businesses in remote provincial towns. They're worse than useless.

You've been getting on well with the official in charge and it looks like he's decided to help you. He already let you select one box from the three. *Then he opened one of the two remaining boxes and showed you the contents* – it contained a provincial licence. Now he's asking you to decide. You can open the box you originally chose, or you can select and open the remaining one. Should you:

- Stick with what you've got – it's the best choice
- It doesn't matter if you change or not
- Change your selection

Think about this for a few minutes and pick only one answer. It's not a very complex situation and it's not a trick question, but make sure you're not overlooking anything. Give it the same level of attention as a typical business decision.

Write down your answer before reading on (Q7-3).

As a seasoned decision-maker, you should have been able to weigh all the factors and give the correct answer. You should also be able to explain your reasoning.

Write down the reasons before reading on (Q7-4).

Now for the third part of this question. What is your level of confidence in the answer? Is it:

- 50% or less (not sure)
- 75% (fairly sure)
- 95% (sure)
- 100% (certain)

If you're not sure (50% or less) go back and work through the question again until you're *at least* 75% sure.

Write down your confidence level before reading on (Q7-5).

Answers

The correct answer is: *Change your selection*. Now, if you selected another answer, you may not believe that. If so, go back and work through the example again. Don't read on unless you're *absolutely* confident that you're right – or you give up!

Was It True?

Still disagree? Here's how it works. When you select the first box, you have a 1/3 chance of selecting the right licence, because there are three boxes but only one good outcome. That means there's a 2/3 probability that the remaining boxes contain the good licence.

When the official opens another box, we *know* it doesn't contain what we're looking for. That makes no difference to the probability of the one you've already selected – it's still 1/3. So, the probability that the other box contains the right licence must now be 2/3.

Convinced?

Did the explanation above convince you? Maybe you still think you're right? Maybe you think the probabilities reset? Let's try again. Suppose the official *didn't* show you what was in the box but instead, *put its contents into the box you didn't select*. So, the box you've selected has one licence and 1/3 chance of being the one you want. The other box now has two licences, which means it has a 2/3 chance of containing the one you want. When you select *that* box (as you should) you'll take out the bad licence (the one the official put in) and throw it away – that's the step the official is doing for you in the original example.

If you're still not convinced, you can read the section on the American game show "Jeopardy" in Gary Smith's excellent book (see Further Reading). The logic is the same – except with goats!

Confidence

That was an important example for two reasons. First, you probably got it wrong and second, you were probably sure (95%) or certain (100%) that you were right. Just suppose that it had been an important decision – one that could have made a significant difference to your organization. You'd have been willing to stand up in front of the board and vigorously defend your (bad) decision!

Could you have got it right? Yes, but you needed to use System 2. You needed a logical decision-making process. You're always being tempted to use System 1 because it's so much easier, but gut decision-making really is for amateurs. It's important to harness some basic mathematical principles to help you understand how the world *really* works.

Linearity

Let's look at another decision-making opportunity. You want to introduce written procedures into your organization. You'd like more information, so you talk to the CEO of a nearby company, where procedures have been used for years. She tells you stories of people being hampered in their jobs because of all the red tape and explains that they're actually in the process of *reducing* procedures. You also discover that there are similar projects underway in five other organizations in the region. When you tell your managers, they're shocked at the news and want you to confirm that you'll *never* introduce procedures. Based *only* on the evidence above (and not on any personal experience you might have), would that be the right decision?

Write down your answer before reading on (Q7-6).

If you said "yes", then you're using a linear model of the world. In other words, you believe this situation can be described by the following diagram:

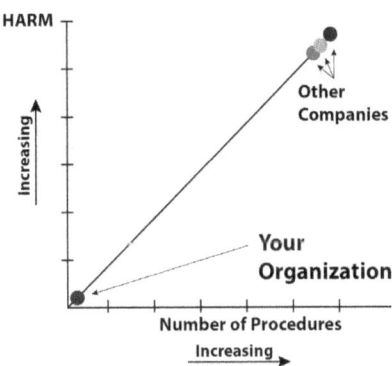

This shows that the organization suffers more harm as the number of procedures increase. You're at the bottom-left with no procedures, so you suffer no harm. The top-right dots are the companies scrambling to reduce the number of procedures they have in place because of the harm they're doing. This diagram confirms that a decision to avoid procedures is correct – doesn't it?

But suppose you're wrong? Suppose the situation is not linear? Suppose the situation is better described by the following diagram:

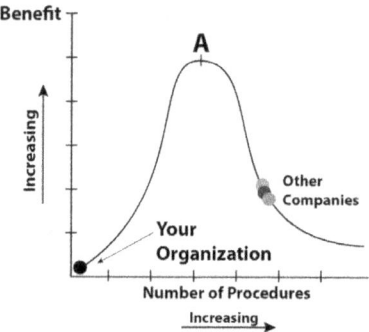

This shows that the benefit you'll get from using procedures will increase until you reach point A. However, if you continue introducing procedures after that, you'll reduce the gain. The more procedures you introduce after point A, the more you're cancelling out the benefits you've got from introducing them.

Now you can see why other companies want to reduce the number of procedures they use. It's not because procedures are bad (the linear model), it's because they've introduced too many. They're still getting some benefits, but these are being offset by new problems. The right decision is for you is to introduce them, but limit the quantity.

Non-Linearly

Many systems are nonlinear. Take your food intake. If you take too few calories, you'll damage your health. As you increase your intake, you'll get stronger and healthier. However, there comes a point where more food starts to damage your body because you're taking too much of it.

In astrophysics, there's a term known as the "*Goldilocks zone*". This is the zone bounded by the minimum and maximum distance from a sun where life (as we know it) can exist. If a planet is too close to the sun it'll be too hot for life, and if it's too far away it'll be too cold. Earth is in the Goldilocks zone in our solar system. You must make sure that key variables in your organization are kept in the Goldilocks zone.

Mental Models

This example demonstrates that if you're using the wrong mental model (for example, a linear instead of a non-linear one), you'll make bad decisions and you'll never realise it. *Humans use mental models for everything* and many System 1 mistakes are due to a mismatch between these models and reality. Graphs and numbers can help reduce these mistakes because they make models explicit.

Caution

The last example also illustrates why you should be cautious about taking lessons from other companies. They may be doing things for reasons that don't apply in your organization. If you don't understand what's really happening, you're likely to jump to wrong conclusions.

Probability and Assumptions

You already have a grasp of *probability*. Everything we experience has some relationship with the concept. Going for a walk? What's the probability it's going to rain? Most people use probability values instinctively. If you hear there's a 70% chance of rain, you bring an umbrella. If it's a 20% chance, you leave it at home. You don't think too much about the details – and most of the time, that's fine. However, sometimes it can lead you astray.

Take a simple case. What's the probability you'll be able to get fresh bread today? Let's suppose you visit the grocery shop on fifty percent of your daily walks. Let's also suppose there's fresh bread on one third of those visits. That means, on average you'll see fresh bread one day in six, so you have a 16.6% chance of getting it today. Would you agree with that calculation?

Write down your answer before reading on (Q7-7).

The math is right, but the result isn't necessarily so. After all, the shop is more likely to be out of bread in the evening rather than the morning. If you walk in the morning, you're likely to get fresh bread *every* day. If you walk in the evening, you mightn't get any. We didn't specify *when* you take your walk or under what *circumstances* you get the fresh bread. The point is that you need to understand the underlying reality – how things work – before you can be sure that the numbers are giving you a correct assessment of the situation. In this example, the numbers were right, but they assumed *a random distribution* and *independence*. In other words, they assumed the following:

- You could take your walk at any (random) time;
- The shop could be out of bread at any (random) time;
- There's no relationship between your walk and the shop's bread status.

Once you state your assumptions, you can see whether they match the situation. If they don't, you need to change your assumptions.

Probability

So, let's look at probability more closely. We can express it as a percentage or a decimal number. If an event has a 20% chance of occurring, it has a probability of 0.2. You can also say it has a 1 chance in 5 of occurring. If something has a 0% chance of happening it means that it *cannot* occur – it's impossible. If something has a 100% chance of occurring, it *must* occur – it's inevitable. And you can't have a probability greater than 100%, so 110% is impossible!

So, if the weather forecast says there is a 20% chance of rain tomorrow, what does that mean? It means that if you had the same set of conditions on a hundred days it would rain on approximately 20 of them. It's *predicting* rain on some occasions. Tomorrow *could* be one of these occasions, but it's more likely to be dry. Now, let's test your instinctive feel for probability.

Suppose you flip a fair coin *four* times...

1. What's the probability that you'll get two heads and two tails? Is it:

- 37.5%
- 50%
- 86.5%
- 100%

2. What's the probability that you'll get either three heads and a tail or three tails and a head?

- 37.5%
- 50%
- 86.5%
- 100%

Write down your answer before reading on (Q7-8).

Did you answer those questions by guessing? It's possible that you didn't take the time to work it out – even though it's relatively easy. All you had to do was construct a table of all the possibilities, then count the number of events you're looking for, and divide by the total.

When you do that, you'll see that there are 16 outcomes in total, and six ways you can get two heads and two tails – that's 37.5%. On the other hand, there are eight ways to get either three heads and a tail or three tails and a head – that's 50%. How did your instinct fare? Many CEOs get it wrong, because their mental models are skewed. Have a look at the following table and confirm that the results above are correct (don't assume – they may not be!).

NUMBER	RESULTS				HEADS	2H + 2T	3H + 1T	1H + 3T
1	T	T	T	T	0			
2	T	T	T	H	1			1
3	T	T	H	T	1			2
4	T	T	H	H	2	1		
5	T	H	T	T	1			3
6	T	H	T	H	2	2		
7	T	H	H	T	2	3		
8	T	H	H	H	3		1	
9	H	T	T	T	1			4
10	H	T	T	H	2	4		
11	H	T	H	T	2	5		
12	H	T	H	H	3		2	
13	H	H	T	T	2	6		
14	H	H	T	H	3		3	
15	H	H	H	T	3		4	
16	H	H	H	H	4			
						6	8	

Real Life

What's the point of all this? In real life you won't be flipping coins, you'll be dealing with far more complex situations. But *that's* the point. If your instinct doesn't give you the right answers with something as simple as coin flips, how do you expect to deal with more complex situations? Probability gives you a way to improve your decision-making skills, but you must know how to avoid the traps.

Let's look at another situation. Your purchasing department has bought two lots of components (A and B) but they've been mixed randomly together. The number of components of each type is exactly equal and they look almost identical. They've hired two inspectors to separate them – but it's a difficult job. The inspectors are taking ten units at a time and separating them. The first inspector has completed four batches and got the following results:

- First ten units: five As and five Bs.
- Second ten units: five As and five Bs.
- Third ten units: five As and five Bs.
- Fourth ten units: five As and five Bs.

The second inspector has also completed four batches:

- First ten units: one A and nine Bs.
- Second ten units: three As and seven Bs.
- Third ten units: three As and seven Bs.
- Fourth ten units: six As and four Bs.

Is that what you'd expect or is something wrong?

Write down your answer before reading on (Q7-9).

Your Decision

You may need to re-train the first inspector. Why? Because the odds of getting exactly 50% in one sample of ten are low (about 25%) so the odds of getting four of these in a row are very low (less than 0.4%). It's very unlikely to happen by chance. It's more likely she knows that you're expecting a 50/50 mix and is unconsciously trying to achieve that proportion.

Random Sampling

If you know the makeup of the population – as you do in this case – you might think that every random sample taken from the population will have those proportions – but it won't. If you took enough samples you'd see *every* combination turning up, including samples with all As and all Bs. *Any* combination is possible – it's just that some will turn up more frequently than others.

The Law of Small Numbers

This also works in reverse. If you select a sample (say ten units from a mixed batch) and you get eight As and two Bs, you might assume that 80% of the components are As. But that's unlikely to be true. Why? *Because the sample size is too small.* We can only have *confidence* in a sample when the sample size is large enough.

This error is called *the law of small numbers* and it can have a devastating effect on decisions. Suppose you've happened to observe a team leader making five excellent decisions in the last year. Now, you want to offer her a manager's job. But stop and think. You've only seen five examples – that's a very small sample. Are you willing to risk hiring the wrong person just on that evidence?

We frequently make decisions based on patterns that could easily be due to chance. We don't take the time to consider the implications of the law of small numbers. Yet the evidence is all around us. Let's say you're in the casino in Monte Carlo watching the roulette wheel. You notice that the ball has landed on black on the last five spins. Is this a good time to put money on red?

You'd think it *must* be – but the answer is no. *The wheel has no memory.* It doesn't remember what the last colour was, so the next selection is as likely to be black as red. You're seeing a random pattern and assuming that it has significance when it doesn't. CEOs make many decisions based on a similar misunderstanding.

Averages

Now you have another decision to make. You've got five work teams all doing the same job in your organization. Three of them are large teams with 100 workers in each, and the other two are small with five people in each. You want to determine if it would be better to break the large teams into smaller units *or* merge the small teams into the larger ones.

You have performance figures for each team but, obviously, there's no point in looking at team totals, because the big teams produce more than the small teams. Instead, you get the team *average* by dividing the team total by the number of workers. The average production per worker on each team from the *first* quarter of 2018 is as follows:

- Team 1 (large) 5045 units
- Team 2 (large) 4986 units
- Team 3 (large) 5034 units
- Team 4 (small) 6233 units
- Team 5 (small) 7117 units

The average person on Team 1 has produced 5045 units while the average person on Team 5 has produced 7117 units. Everyone has been properly trained and there's no reason to think it's due to skill level. The production manager is sure the difference is due to the team size. The smaller teams have more team spirit than the big ones. His vote is to break up the big teams. Do you think that's a good idea or could there be some other reason for the difference?

We need to look at the sample size again, but this time for a different reason. There's a hint in the numbers. Did you notice that the average figures for the three large teams are much closer together than those for the smaller ones? Let's have a look at the figures for the *second* quarter of 2018:

- Team 1 (large) 5011 units
- Team 2 (large) 4992 units
- Team 3 (large) 4945 units
- Team 4 (small) 4352 units
- Team 5 (small) 6644 units

The averages for the three larger teams are close together again, and similar to the first set. But the small teams are showing big differences. Why? *Because the average from a small sample size is likely to be more extreme than that from a large sample* – it's another way of looking at the law of small numbers.

Think of it this way: you're likely to get variations across people in every team. Some are excellent workers, most are average, some are not so good. People will have good days and bad days, and some will be off sick. Individual outputs will vary considerably for many reasons. However, in large teams, much of that variation will cancel out. You're also dividing by a bigger number (100 in our case) and this will tend to reduce the size of any remaining variation. In small teams, you're less likely to have the variation cancelling out and you're dividing by a much smaller number (5 in this example). That's why you can expect more extreme results from small teams. You see this phenomenon everywhere:

- The average output per worker will be much better or worse for smaller organizations.
- The average exam results will be much better or worse for small schools.
- The number of deaths per million for a rare disease will be much better or worse in smaller countries.

This issue affects many calculations. For example, if you're comparing the average daily output from workers but you use three years' data for some but only six months' data for others, you'd expect to see either a higher *or* lower result for those with less data. This does not (necessarily) mean they are better or worse than the average.

The important point here is that *the way* you select data can be responsible for giving a false impression. The numbers could mislead you into thinking that something (a person, team, company, etc.) is better or worse than it is. It takes System 2 thinking to avoid falling into that trap.

Data Mining

You've probably heard of *data mining*. It means searching through massive amounts of data, looking for patterns that can be used to make decisions. It can be used to expose hidden relationships between seemingly independent variables. It has the potential to be extremely powerful.

Let's say the data mining manager comes to you and he's excited. He's found a correlation between the FTSE 250 Index and sales of your product to male customers between 20 and 40 years old. Even better, the correlation is offset by three weeks. In other words, when the FTSE goes up, sales of the product go up three weeks later. That lead time will allow your manufacturing team to ramp up to meet demand. You ask him how sure he is, and he tells you that he's applied a statistical test and there's only one chance in a million this could be due to chance – it's solid. What do you think?

Write down your answer before reading on (Q7-10).

Data Patterns

Before we look at the answer, let's think about patterns in data. Suppose you have a coin that you believe *isn't* "fair". You think it's been tampered with and you want to prove it. You flip it four times and it comes up heads each time. Is that proof it's not fair?

If you look at the table earlier in the chapter, you'll see that there's only one chance in sixteen of getting four heads. So, it's unlikely that would happen by chance – but it's still possible.

So, you continued to flip it until you've got seven heads in a row. Now you're a lot more confident that something is wrong. The odds of getting seven heads in a row by chance are less than 1%. But you're still not satisfied and continued until you get ten heads in a row. Now you're very confident, because there's only one chance in 1024 of getting that result by chance.

Now, let's look at an alternative situation. Suppose I've been given the results of *millions* of coin flips from thousands of coins under different conditions. The data contains additional information like the date and time, and the sex, age and height of the flipper. I search through all the records until I spot a run of ten heads for a specific coin. However, there were also some tails in the results for that coin. I analyse the results further and eventually find a pattern. I announce that the coin was *not* fair (all heads) when flipped between 3.00pm and 5.00pm by a flipper aged between 25 and 30. I calculate the odds of this happening by chance as one in one billion. Is that result as valid as you flipping your coin?

Of course not. The odds of the result (ten heads) are exactly 1. *It's already happened.* Don't forget, I'm likely to find one such run in every 1024 results and I've got *millions* of results. All I did was find a way to make the outcome look more significant than it was. I selected the results I wanted by filtering.

Back to Data Mining

It's probable that your data mining manager has done the same thing. He's searched for patterns and found one. He's filtered the data (male customers between 20 and 40) to make it fit. Also, he isn't just looking at current data – his matching pattern is offset by three weeks. That's suspicious. The likelihood is that *this is just a random pattern.* There are several flags that should make you sceptical in a case like this:

- Someone is trawling for *any* pattern rather than a specific one;
- There's no theory to account for the results
  (why is the FTSE related to your sales?);
- Some of the data has been removed (filtering);
- The same data is being used (a) to find the pattern and (b) to prove its significance;
- The significance test is misleading.

So what good is data mining if you can't depend on its results? You *can* use it to find patterns but then you *must* create a theory to explain those patterns and *make a prediction* based on that theory. Then – most importantly – you must *test* the prediction with new data. It's very important that new data is used to test these predictions. Yes, you could use two sets of old data – the first (say last year's) to find the pattern and the second (this year's) to prove it. However, you can't be certain that someone hasn't already used the second set in creating the prediction.

Significance Testing

Medicine has had a chequered history. For most of recorded history, the "cures" it offered were ineffective at best and highly dangerous at worst. Schoolchildren were advised to smoke cigarettes (as a disinfectant), doctors used radium to treat diabetes and rheumatism, and bloodletting was common! That situation continued into the early decades of the twentieth century. It was only then that a technique, developed by the statistician Ronald Fisher, began to be taken seriously by the medical establishment. It was called *significance testing* and a version is used today to confirm the effectiveness of new treatments.

How does it work? Let's say I have a new potion that I claim will cure hay fever. Prior to 1900, that claim was all I needed to sell it as a miracle cure. Today I need to prove that it actually does what I say. You might think that's easy. Just give it to someone with hay fever and see if it cures them. However, it's a bit more complicated than that. If I give it to someone and they appear to get better, is that due to the potion or would they have got better anyway? Maybe any apparent benefit is due to the *placebo effect?* This is where a patient thinks she's getting better because she *believes* the potion is helping her, even though it has no medicinal value.

The Null Hypothesis

We need to eliminate these possibilities, so instead of assuming that the potion works, we start by assuming that the potion *doesn't* work and any positive results we're seeing are due to chance. This is called the *null hypothesis.*

Controlled Experiments

Now we must test that assumption. We do that by setting up a *controlled experiment.* We randomly assign volunteers with hay fever to two groups. The randomness of that allocation is important. Then we randomly assign one group to take potion A and the other to take potion B. One of the potions is just a harmless chemical with no medicinal value – a placebo. The other is the test potion. Neither the volunteers nor the doctors who administer the potions and track the results know which the real one is. This is called a *double-blind test*. That's important, because people tend to unconsciously bias the result if they know the desired outcome.

Now the test is run. Efforts are made to eliminate any confounding variables, like the time of day or the location where the test takes place. If it's not possible to eliminate these variables, they must be assigned randomly. Then the results are analysed. If the null hypothesis is correct, we'd expect no significant difference between the two groups. Of course, there will be *some* difference. You might find, for example, that the daily number of sneezing fits is 12% lower in the potion group than it is in the placebo group. That looks good, but is it significant or just due to random variation?

For a result to be accepted as significant in medical research, there *must be less than a 5% chance that it could happen by chance*. We won't do the calculation here, but the principle of significant testing is to measure the *inherent variability* within each group and compare that to the differences between the groups. Suppose the sneezing behaviour of the test subjects was very erratic – sometimes they'd sneeze a hundred times a day, sometimes only ten or twenty times. In that case, we'd need to see a *huge* difference to establish significance. On the other hand, if they all sneezed between a hundred times and a hundred and five times per day (reasonably regularly) then a much smaller difference would be significant.

From Medicine to Business

Why this foray into medical history and statistics? Because medicine didn't start to make real progress until it began to embrace scientific methodology. Up to that point, *everyone had an opinion*. Doctors and patients listened to "experts" who knew as little as they did. There was no point of reference – no hard facts. It's the same in business today. In many organizations, the "truth" is simply *what the most important people believe*. Most are CEOs and managers, but you can also find accountants, engineers, programmers and administrators who spread their own gospel – their model of the world. And if nobody challenges them, they grow more dogmatic. It's never good to accept unassailable authorities in any organization. They're almost certainly wrong about many things and the damage they can do is immense. These people are absolutely convinced they're right and will get emotional when challenged. They'll also be quick to claim they have evidence for their beliefs. That "evidence" needs to be probed and tested. Much of it will be based on personal experience. It will take the form of anecdotes and stories. And it *will* suffer from *the law of small numbers*. But how do you explain that to people without the necessary background? We'll come back to that later.

Regression to the Mean

You're thinking of expanding overseas. You've identified three companies that might be worthwhile partners, so you've been reviewing their performance over the last five years. Up to last year, they were all achieving an average of around a 10% return on assets. Last year the results were as follows:

- Company A: 10.1%
- Company B: 10.3%
- Company C: 17.2%

Company C had a terrific year. They've obviously been doing something right, so you're thinking of approaching them. However, the chairman of your board wants you to make a prediction about their results for this year. What would it be?

Write down your answer before reading on (Q7-11).

In the absence of any further data – you should expect their return this year to revert to around 10%. Why? The reason is known as "*regression to the mean*". It basically says that if you see an abnormally high or low reading, it's likely to be an outlier. The next result is likely to be closer to the mean. Consider possible reasons that the company did well last year. It might have received an unusually large order or perhaps it was paid up front for products that will be delivered in the new year. In other words, they were lucky because some chance event occurred but it's unlikely that the same will happen again this year. Of course, they *may* have done something that fundamentally improved results but, in the absence of hard data, it's safer to assume that regression to the mean will take place.

Big Decisions

You still want to expand – but now you're looking at a completely new market. There are five companies competing in this market and it looks like they're all performing well. After a detailed investigation, you find that the average annual increase in sales for the last five years was an amazing 31%. That figure looks very attractive but you're being careful – are you overlooking anything?

Write down your answer before reading on (Q7-12).

Survivors

It's important to remember that the companies you're looking at are all in business *now*. What about companies that were in business five years ago but are no longer around? Suppose there were thirty companies then and that's been reduced to five? When you take their figures into account, you find that the average increase in sales over five years is less than one percent. Why? Because the market is relatively static and the survivors achieved their increase in sales by grabbing customers from companies going out of business. So, the five companies that exist right now have gobbled up all their competitors – and may do the same to you if you try to muscle in.

When you limited your analysis to existing companies and ignored companies that *previously* existed, you were fooled by *survivor bias*. It's easy to do. You must *force* yourself to ask if relevant samples are missing when considering any data. This is best illustrated by the story of the statistician asked to determine the best way to armour US bombers during World War II. The Air Force had already collected data and found that most of the bullet holes were in the fuselage. They were ready to attach armour to those areas when the statistician told them to stop. Can you figure out why?

Write down your answer before reading on (Q7-13).

Isn't it obvious? He pointed out that all the data had come from planes that had made it back. He suggested that they should add armour to the places where the bullet holes weren't. The reason? The planes hit in those areas never made it back! Yes, it's obvious when you think about it – but you *do* need to think about it. Gut reactions are dangerous because they *prevent* you thinking.

FURTHER READING

Smith, Gary, *Standard Deviations: Flawed Assumptions, Tortured Data and Other Ways to Lie with Statistics*, Gerald Duckworth & Co.

8 METRICS

You probably monitor several key metrics: (units shipped, hours worked, equipment utilization, quality indicators) that tell you how well your business is doing. You watch them every day for any indication that things are slipping. If you spot any anomalies, you immediately ask questions and make sure the relevant team leader acts. But what's the effect of these metrics on employees?

Imagine you run a technical support line and you're monitoring the call response time. This is measured from the second a call is received to the second it's answered. Team leaders are pushing their teams to beat the target as they approach the end of the month. If you're an engineer and you can't answer a call within the target time, what do you do? Easy: *you don't answer it at all*. Since the call goes unanswered, it doesn't drag down your metrics. Of course, that's not great for the customer but when you're under pressure, that's not your main consideration.

That's what's called *metric-induced dysfunctional behaviour*. The attention placed on metrics and goals produces undesirable behaviour. It's particularly common where targets are felt to be unreasonable and can take many forms.

Let's say you're measuring the time it takes the engineers to *resolve* an issue. If they're struggling, they might give sub-optimal solutions to customers just to "resolve" it quickly. For example, they might tell them (untruthfully) that they're suffering from a software bug that'll be fixed in the next release. They could also transfer potentially difficult calls to other engineers, ask the customer to call back or tell them they'll send the solution in an email (and never do). They might even transfer calls to other departments. Unscrupulous people will exploit any weaknesses in the system.

When that happens, honest people find themselves at a disadvantage and resent both the people who take advantage *and* the organization itself for not addressing the problem.

Preventing Dysfunctional Behaviour

The more knowledge you have about a process, the less likely that dysfunctional behaviour will become a problem. Let's say you want the average resolution time to drop from 7 minutes to 5 minutes. *Before* setting that as a target, you should understand exactly what's happening – and the best way to do that is by taking measurements and analysing the results. Let's say you found the following:

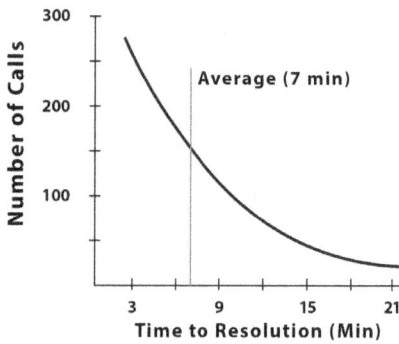

What does that tell you? Well, most of the calls last less than 7 minutes. The problem is that the average is being pulled up by a relatively small number of calls. To illustrate, suppose we have five calls:

- 3 minutes
- 4 minutes
- 5 minutes
- 3 minutes
- 20 minutes

The average is 7 minutes. You could work on reducing the first four, but there's not much to be gained. Let's say you reduce each by 20%. That gives you an average of 6.4 – an improvement, but you haven't reached the target. However, the fifth figure is relatively large so if you could reduce *just* that one by 50%, your average would drop to 5.

How could you do that? Well, you need to gather additional information. You need to find out *why* some calls are taking so long. You could get the engineers to identify the issues. Perhaps these calls deal with more complex issues, or the customers are more difficult. The point is to use *data* to pinpoint the exact problem and then come up with options to tackle it – *don't just set an arbitrary goal*. Metrics influence behaviour but not always in the way that was intended. That's why it's important to consider them carefully. Sometimes they mislead and actually hide what's *really* going on and that could lead to bad decisions and negative consequences.

Proxy Metrics

It's not always possible to measure things directly, so you might select a substitute. For example, you might want to understand how satisfied your customers are feeling. You can't do that directly, but you could measure complaints and assume that the lower the number, the better the service. You're using the number of complaints as a *proxy* for what you really want to know.

If you use proxy metrics to manage your business, you risk introducing problems. Can you think of any reason why measuring complaints might not be a good indicator of customer satisfaction? That's right: because you're assuming people will make the effort to report problems. Some people will, of course, but many won't. You might get 20 complaints in a week but there could be hundreds of dissatisfied customers you know nothing about.

You could argue that the absolute number isn't important – it's the *relative* number that counts. If the number is going up, the service is getting worse, and if it's going down, it's getting better. Does that make sense, or have you forgotten anything?

Write down your answer before reading on (Q8-1).

You may have forgotten about *the law of small numbers*. Even if the quality of your service remains the same, you're likely to find differences in the level of weekly complaints due to random factors. For example, some weeks, just by chance, you'll get more people who're inclined to complain. What happens when a manager notices an increase in the number of complaints? She'll want to act. She might berate the team leaders or threaten the workers with the loss of their bonuses. Of course, they're likely to keep on doing what they've always been doing (System 1s in action). The following week, the numbers are back to normal. The manager is convinced that *her* intervention has made all the difference, so she'll continue using that behaviour – she's "proven" that it works.

CEOs sometimes use proxy metrics without even noticing it. They reward managers who work late. That's rewarding the proxy (time spent in the office) rather than the results (higher productivity, less waste). Review your formal and informal metrics and try to eliminate as many proxies as you can – they'll only lead to discontent if less-scrupled employees are able to game them.

Pricing and Costs

Let's move on to another area where numbers and behaviour are intertwined. The price you set for your product or service can have a significant impact on sales and profitability. Cost is an important factor in setting the right price – at a bare minimum, you don't want to make a loss. However, CEOs and marketing managers have been known to make "one-off" decisions that ignore cost, in order to get short-term sales. Then they forget about those decisions until the negative impact on profits is revealed.

What would you do if you were faced with the following situation? Your salespeople have identified a new customer with great potential. They tell you that you'll get the initial sale if you're willing to lower the price. The cost structure is as follows:

- Materials: €1.23
- Processing: €2.35
- Overhead: €5.44
- Profit: €1.33
- Price: €10.35

They want to place an order for 12,000 units at €8.95. Technically that's a loss but the salespeople assure you that the customer has huge potential and this order will get you in the door. They point out that more than half the overhead is administration costs, and that could be allocated anywhere.

Since overhead is the single biggest cost, it's tempting to make a few adjustments so that more of the burden is carried by other products. It wouldn't be forever – it's just to land the customer and get more sales. After all, the quantity is relatively small so there'll be no *actual* increase in overhead cost. And since the cost is already covered by existing products, you could even afford to drop it completely on this order. Does that make sense?

Write down your answer before reading on (Q8-2).

Salespeople often argue that profits are coming on future orders. However, if they've not been successful in getting a good price on the first order, *it's unlikely they'll succeed when the quantities are higher*, particularly if they're being paid a percentage of total sales! And while overhead costs may *seem* flexible, changing them can have serious consequences if the implications are not monitored.

Allocating Overhead

It's relatively easy to calculate the cost of the materials and direct labour that goes into a product. But how do you allocate overhead like management and administration costs? How do you factor in the depreciation of machines and equipment? What about the cost of the support departments like maintenance and stores? One way to do it is to assign a blanket overhead rate. In other words, add up all the overhead and then allocate this based on the processing hours for each product or service. For example, assuming it takes one hour to make each unit:

- Total overhead cost: €510,000 / year
- Total processing time: 6000 hours
- Overhead on each unit: €85

That's easy enough when dealing with a single product but it can lead to problems when there are multiple products with different operations. It may even make some products uncompetitive.

The Two-Stage Overhead Process

Another way to do it is called the two-stage overhead process. This uses the concept of "*cost centres*". For example, an assembly operation might be in one cost centre and stores might be in another. There are four steps in the process:

1. Allocate overhead to production and support cost centres;
2. Re-allocate support centre costs to production cost centres;
3. Calculate separate overhead rates for each cost centre;
4. Assign overhead cost centres to products.

Here's a simple example:

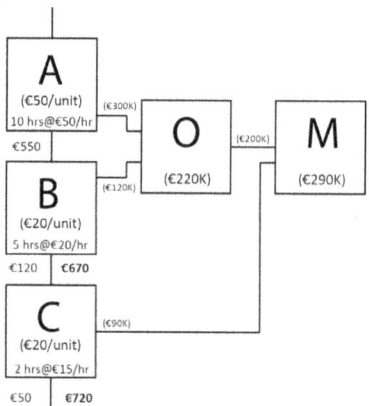

- A, B and C are production cost centres
- M is management and administration overhead (€290K/year).
- O is "other" overhead costs (€220K per year).

The administration costs (M) are allocated to O (€200K) and to C (€90K). The total O costs (€200K + €220K) are then allocated to A (€300K) and B (€120K).

The overhead cost in each production cost centre is allocated to the product based on the processing time. For example, in cost centre A, each unit takes 10 hours to produce and there is €50 of direct cost.

The total cost allocated to each product after cost centre A is €550, of which €500 is overhead. Likewise, the total cost for each product after B is the direct and overhead cost for that cost centre (€120) added to the cost from A (€550), giving €670. The total cost after C is €720. If you were to add a profit of 20%, the selling price would be €864. Based on those figures, if you sell 810 units you'll make a profit of €116,640.

New Product

Now, you want to introduce a new product. This one has the following cost structure (using the same overhead rates):

- Cost centre A : €100 + 5 hrs @€50 = €350
- Cost centre B: €120 + 2 hrs @€20 = €160
- Cost centre C: €60 + 10 hrs @€15 = €210
- Total cost: €720
- Selling price: (+20%) = €864

The total processing time is the same (17 hours) in both cases, and so is the total cost (€720) and the selling price (€864). You should be able to replace one product with the other with no difference to your profit... or are you missing something?

> **Write down the answer before reading on (Q8-3).**

The total processing time *is* the same (17 hours) in both cases but the direct costs in the second case are higher and there's less going to overhead. That's because we've reduced the processing time in two areas, but we've increased it in the third and the overhead allocation is lower in that area. If we were to replace the first product with this one and sell the same number of units, we'd be in trouble. In the first case, our overhead was covered, and we made a profit when we sold 660 units. In the second case, there's a shortfall at that volume – we're losing money.

Of course, it's unlikely that managers would overlook something this obvious, particularly when there's only one product. In real life, however, where there are many products and cost centres, it could easily happen that a specific product is being sold at a loss. If you're not already familiar with the overhead allocation system in your organization, you should do a little digging and then discuss the implications with your management accountant. The results of getting it wrong can be disastrous.

Quick Review

Did you work through that example? Do you understand *exactly* what was going on at each stage? If you skipped over it, then you've confirmed again that your System 2 is lazy – too lazy to get involved in an important area that could impact your organization. Perhaps you *did* pay attention? Either way, can you answer these questions (*without looking back*):

- Was there a calculational error in the example?
- Is the answer wrong by more than a factor of 2?

Write down the answer before reading on (Q8-4).

If you can't answer those questions, you didn't pay enough attention to the numbers. Remember, there was *no* complicated maths in this example; it was just addition, subtraction, multiplication and division. However, it *did* take a bit of effort and concentration – it needed System 2 involvement. If you skipped over the detail, it's likely that *you do the same in other situations*.

Perhaps you glaze over when you're being briefed by your financial controller or management accountant? That's dangerous because you've got no way of knowing if they're telling you the truth if you don't engage. Could *their* figures be off by 10%, by 50% or even by 200%? How can you be sure?

By the way, if you're thinking that this example doesn't count because it's not *really* important – or if you're thinking that it's really OK you've skipped over it but you'll get back to it later – you're in *denial*. You're never going to do it. That's just your brain shielding you from blame and re-writing history to keep you happy. If you're not willing to accept these negative realities, you're not likely to do anything to address them.

Allocating overhead is just one example of a boring but important job that all World Class CEOs need to confront.

Your Metrics

It's coming to the end of the financial year and you're going to miss your sales targets. The shareholders won't be happy, and you and the other managers won't get your bonuses. Suddenly your sales manager comes up with a solution. If you were to drop prices by 25%, he'd be able to persuade some of the bigger customers to bring their orders forward and that *will* bring you in on target. Should you do it?

> **Write down the answer before reading on (Q8-5).**

Let's consider why you might say "yes". First, you'll meet your targets and the shareholders will be happy. Second, you'll satisfy the managers, and they'll be happy. And, of course, you'd also get your own bonus (as if that was a factor!) and look good.

But look at the negative side. You're giving away a sizeable slice of money and, in the process, reducing short-term profits. You're also negatively affecting long-term profits because your customers' purchasing departments will latch onto the incident to negotiate better prices at your expense. Bottom line? You've reduced shareholder value.

You're also degrading the culture of the organisation. Employees *will* find out what you did and convince themselves that you and your cronies are lining your pockets at the expense of the organization. They won't make any sacrifices after seeing your attitude.

But what about the shareholders? If you don't meet your sales targets, they could lose confidence and the share price could drop. Isn't that a valid reason to say "yes"? Of course not, that's just System 2 justification. The share price will go up and down for many reasons, most of them outside your control. One of your jobs is to manage expectations. You must let the board and shareholders know that you're *not* willing to make short-sighted decisions to artificially maintain the share price. That *could* mean it'll drop in the short term, but value investors will realize that your approach will give them long-term value. World

Class leaders won't be tempted to make a decision that will damage the organisation to achieve an artificial goal.

Motivation

The fundamental topic in this chapter was *motivation*. How do you motivate your employees to work as hard as possible and make the right decisions? How do you motivate them to avoid bad behaviour? Most performance management systems lead to negative consequences because they're too simplistic. They focus on a limited number of outcomes and disregard any fallout.

It's easy to believe you know how to motivate people, that there's a relatively simple solution, but that's System 1 thinking. The reality is different. Before you can design an effective reward system, you need to have an effective *decision-making process*, one that will look beyond the first-order effects of each decision. We'll be looking at such a process in a later chapter.

FURTHER READING

Drury, Colin, *Management & Cost Accounting*, Chapman & Hall.

9 FORECASTING

Every time you make a decision, you're also making two predictions about the future. First, you're predicting what will happen if *no* decision is made, and second, what will *change* once the decision is executed. For example, a manager believes that the sales department's productivity will drop without a team leader. When she hires a new person for the job, she thinks he'll successfully increase productivity and will contribute more than his salary to the organization.

Usually, you don't even think about these predictions – you simply make the decision because it seems the right thing to do. Nevertheless, there's always a prediction in the background. Even when you make a mundane decision like what you'll eat for lunch, you're predicting that you'll feel hungry and a sandwich will make you feel better.

Many management decisions are sub-optimal or plain wrong. There are several reasons for this, but the one we're going to tackle in this chapter is the failure to forecast correctly. Of course, predicting the future is never easy, there are many confounding factors, but *lack of effort* is a significant cause.

Remember those sandwiches you bought on your way to work? When you arrived at your desk, you remembered that you're taking a client to lunch, so the sandwiches are going to be wasted. That's a failure to predict the future, even though you already had all the facts. It's hard to predict the future if you don't give it any thought! It's better if you take a systematic approach and in this chapter, we'll look at some tools you can use to do that.

Sales Forecasting

Most organizations attempt to predict the future at least once a year. They try to predict how many support calls they'll need to handle, or how many smartphones they'll sell, or how many haircuts they'll give.

They need to do this to budget for next year, and it's important because it informs the decision-making process.

For example, if you predict that you're going to sell 50% more smartphones next year, you'll need more equipment and people to manufacture them as well as more sales and support staff. That means you'll have to borrow money, buy and commission new equipment, hire and train new people, perhaps even expand into new premises. There's a lot riding on that prediction. Of course, you could postpone it and just wait to see what happens, but then you risk losing sales and your competitors will be quick to take advantage of your inaction. So, how much credibility do you give to the sales predictions in *your* organization?

Write your answer as a percentage (Q9-1).

Your answer probably depends on how the predictions were created. Do any of the following look familiar?

- The salespeople came up with the predictions.
- It's based on last year's sales.
- It's based on years of industry experience.

In the first case, you probably don't know *how* the figures were arrived at; it could just be a guess. In the second, you're assuming that the future will be the same as the past. The third is just a different way of saying "gut feeling". None of these is good enough, yet many companies use them every year!

Working the System

Why do CEOs use sub-optimal systems in such an important area? Many will respond with: "Our results have been acceptable, so there's no need for major change." They believe that historical results are enough. Let's put that belief to the test.

Suppose a successful businessman gives up working and, instead, buys a large quantity of lottery tickets every week. He wins only small amounts for the first six months but on week 32, he wins the jackpot of €130 million. Did he make the right decision? Was he right to give up his job and bet on the lottery?

Write down your answer before reading on (Q9-2).

You know the right answer, but you might have hesitated – after all, he *did* win. But you *should* remember that the odds of winning the jackpot are over 100 million to one. He happened to get lucky, but it was far more likely he'd lose his money. There's another way to look at it: if he asked for your advice *before* he implemented his "strategy", what would you have said? It's likely that you'd have told him not to be stupid because you know how it should have turned out. This points to an important rule:

You can't judge a process or system by results alone.

Of course, many CEOs disagree with that principle. They'll tell you that results are the *only* thing that matter. They'll argue that it *doesn't matter how you get there*, as long as you get the results you want. So, let's pretend the following happened to you: you get a letter in the mail from a stockbroker telling you that they've created a new AI program to predict the stock market and as proof, they identify a stock that's going to go up in the next few weeks.

You watch out of interest and, sure enough, it does. You get another letter the following week, telling you about a stock that's going to go down. You watch, and it also happens – maybe these people know what they are doing? The same thing happens again and again and again. At last, they offer you the opportunity to invest with them. They've given you plenty of evidence that their system works – so if you *really* believe in results, you can't go wrong. Are you ready to invest your savings with them?

Write down your answer before reading on (Q9-3).

They've given you evidence that the system works… but of course, it doesn't really – it's a scam. It works by sending out (say) 2000 letters at the start. Half of those letters announce that the stock will go up, the other half that it will go down. Then, the stock will either go up or down. The incorrect half will be abandoned, and the correct half will get a second letter with the new prediction. This happens again and again until there are only 125 left, all of whom received five correct predictions – and you were one of the "lucky" ones.

It was just the luck of the draw that *you* were one of the lucky ones – the odds were against it (16 to 1) – but *you only saw the end result*, so it was natural to assume that they got their forecast 100% right. The system *seemed* to work. But just because a system *seems* to work doesn't mean it does Just because a CEO *appears* to make four or five good decisions in a row doesn't mean she'll continue to get good results.

Predicting Sales

Many CEOs use last year's sales as an anchor and predict a relatively small increase. That'll work reasonably well until a big change occurs – particularly a big *drop* in sales. That'll catch them off-guard and the results are likely to be serious. That illustrates an important point. When you're predicting, it's often not the *usual* that's most important – *it's the unusual.*

If you experience four or five years in a row where there's been a small increase in sales, you're likely to believe that trend will continue. You might become so confident that you dismiss any attempts to develop a more accurate system. Your bias will affect others, so nobody will be looking for the warning signs of a change. When it does occur (and it will) you'll be caught unprepared and end up struggling. But is it possible to create a more dependable predictive system?

Monkeys With Darts

You probably read about a 2005 study on *expert judgement*. It was reported that *the average expert is about as accurate as a dart-throwing chimpanzee!* The researcher who carried out that study was Philip Tetlock and many pundits used his results to conclude that predicting *anything* with any certainty is impossible – even by experts. As usual, the actual results were a bit more nuanced. In fact, Philip describes himself as a "forecasting optimistic sceptic". He worked with the Intelligence Advanced Research Projects Activity (IARPA), a US government agency, on the Good Judgement Project from 2011 to 2015. Over that time, his group of forecasters – ordinary people without specific expertise in the areas they were predicting – outperformed all other groups, including professional intelligence analysts with access to classified data.

Forecasting Guidelines

Philip has come up with a set of simple rules that can help anyone become a better forecaster. *You* can use them to improve forecasting in your organization. Before we look at the rules, do you think it's going to be worth the trouble of using them? Is it important to get more accurate predictions in *your* business?

Write down your answer before reading on (Q9-4).

Hopefully, you said "yes", so here are the key rules:

1. Work with others;
2. Break the problem down into its constituent parts;
3. Look for opposing causal forces;
4. Balance the inside view with the outside one;
5. Use numerical values to estimate probability;
6. Balance between overreacting and underreacting to evidence;
7. Balance under- and over-confidence;
8. Implement with feedback.

1. Work with others

Get many people involved in the forecasting process. Each will have a different background, a different outlook and will react differently to the collective knowledge of the group. The accuracy of the forecast will improve when all this information is synthesised.

2. Break the question down

For example, how can you break down next year's sales forecast? Let's start with what you know:

- Where did your sales come from last year?
- Who were your biggest customers?
- Are they going to buy from you again next year?
- How do you know?

If you're dealing with relatively few customers, you can get your sales-people to talk to them individually and find out. If you're dealing with retail customers, you could break them into segments (for example, by sex, age or dress sense) and conduct a survey on a sample of each segment.

That's a start – but you also need to look at other factors: how many new customers are you going to get and how much are they likely to buy? What's your assumption based on? What about the negative side? How many customers are you likely to lose, and how many lost sales? You should also look at your actions during the year. Did you run any promotions? If so, what were the results? Did you drop or increase your prices? Did anything happen to depress or stimulate sales? Could it happen again?

3. Look for opposing causal forces

Now look at the bigger picture. What factors are likely to affect your customers? For example, would a recession affect your sales (and if so, what's the probability of a recession over the next 12 months)? What about an increase in VAT? A trade agreement with China? A significant increase in fuel prices? Identify the direct and indirect factors that will affect your sales and then estimate the probability they'll occur and the likely effect. Some will tend to increase sales, while some will work in the opposite direction. You can also use *Michael Porter's five forces* to identify threats and opportunities:

- Substitute products or services
- Competitors
- New entrants
- Suppliers
- Customers

If you're selling ice-cream, there are *substitutes* like chocolate, yoghurt or buns. If you're selling smart TVs, possible substitutes include laptops, tablets or smartphones. What substitutes are available for your product or service? Are they gaining or losing popularity? Are there important trends that could affect your business? And could *your* product be a substitute in another industry? You should have intelligence on your existing *competitors* but are they likely to do something new? Are they ready to bring out a new product or cut prices? How do you know?

New entrants can be dangerous, particularly if they're using new technology to leapfrog over your traditional approach. If you take too long to respond, you'll be outmanoeuvred and could lose market share. Are there any companies out there – even small ones – with new technologies? What about companies in other industries who might be able to expand their capability to compete with you?

Strong *suppliers* could increase costs and force you to raise your prices, thereby reducing sales. They could also limit how much they'll sell you. Strong *customers* could force you to lower prices. They might even force you to supply them with more product than you want to – reducing your profit margin even further. Are your suppliers and customers likely to be stronger in the future? What factors could contribute to that situation?

4. Balance the inside view with the outside one

You have an inside view of your organization. You know the technologies, the suppliers, the customers. But you may be *too* close. It can be useful to get an outside view. Question people who're not in your organization. Use them to give you another perspective on the industry and the factors that could affect future sales. When you want to estimate how likely something is to happen, look at external base rates. For example, what was the success rate in adapting that new technology?

5. Use numerical values to estimate probability

You've assembled all the relevant information. Now you need to estimate the probability of these events happening. You *could* use vague words like "unlikely" and "very likely", but people will have different interpretations of these terms, so you *need to be precise*. Use numbers to estimate probability.

Numbers are good, but it's important that everyone understands what they mean. For example, if a forecaster said there was a 90% chance of rain today and it didn't rain, was she wrong? Not necessarily. We've seen that the best way to think of the prediction is this: if the situation was to repeat 100 times, it would rain on 90 days and it would be dry on 10 days. Remember, if the 90% is accurate, you *will* have dry days. Now, let's apply this to the sales example. Let's say your sales last year were €5 million. If there were no major changes, you'd expect an increase of 10% this year. So your initial estimate would be €5.5 million. However, you've identified a significant factor that could change that prediction: your biggest competitor is likely to release a new product. You estimate there's a 60% of that happening and, if it does, it will reduce sales by 50%. So, your prediction is as follows:

- There's a 60% probability that sales will be €2.75 million.
- There's a 40% probability that sales will be €5.5 million.

That's a significant difference and it has implications. If the rival product materializes, it will mean cutbacks, re-allocation of resources and, perhaps, a change of strategy. The sooner you begin to address these issues, the more likely you'll be able to minimize the negative effects. However, if you act and it *doesn't* happen, you're likely to damage the organization.

What would you do if you were presented with this situation? Most CEOs will stick with the *status quo*. They'll do nothing, even though the probability is in favour of acting. This raises an important question. What is the *lowest probability* that would prompt you to act? Would you act at 75%? Maybe 90%? Or would you wait until you're 100% sure – even though that means you're being reactive? By then it's probably too late to do anything, and the consequences could be serious.

6. Balance between underreacting and overreacting

In many organizations, forecasting is only carried out once a year. There may be slight adjustments to the figures based on the previous month's sales, but no serious attempt is made to review the original forecast.

In contrast, Philips's forecasters continued to *actively* search for new information and adjusted their forecasts when relevant data were found. That means the probabilities were continually changing. That approach contributes to better forecasting results, because the underlying reasons for the forecasts are being repeatedly challenged and revised.

However, this process can result in two types of error. The first is *underreacting to evidence.* The original prediction acts as an anchor and all the later evidence results in relatively small adjustments to that figure. It's important to look at each piece of evidence and try to determine its significance without being *unduly* influenced by what has gone before. The second error is the opposite of the first – it's *overreacting to evidence.* This is where a new piece of evidence comes in and results in a big shift in the probability. It's important to carefully review this evidence and see why and how it conflicts with what was already known.

7. Balance under- and over-confidence

If you're *under-confident* in a forecast, you're likely to ignore it when making decisions – even though it's the best estimate of the future that's currently available. By ignoring it, you're assuming that another outcome is more likely – usually the status quo.

If you're *overconfident* in a forecast, you're likely to ignore external changes and other relevant factors. You're likely to rely *too* heavily on the initial prediction when making decisions. This can happen if you haven't been deeply involved in the forecasting process.

8. Implement with feedback

One of the problems with many prediction systems is that they're open-ended. A prediction is made but there's no review afterwards – no attempt to check if the prediction came true or not. No chance to identify what worked and what didn't.

Forecasting is a process. We've seen that it's best to continually adjust the prediction over the course of the year – actively looking for new information and adjusting the probabilities carefully as required. It's also important to review what happens and find new ways to improve the process.

At the end of each forecasting period, it's time to look back and see what happened. Here again, it's beneficial to take a structured approach by asking a series of questions. For example:

- What led people to make the original predictions?
- Were those factors all equally valid?
- What was ignored or overlooked?
- How could the process be improved?
- How did the predictions change over the year?
- Did people overreact or underreact to evidence?
- Was evidence ignored or overlooked?
- Could the updating process be improved?

You can see why it's necessary to document each piece of evidence, each decision and every one of the predictions, and track them as they change. This is the raw data that'll ensure your prediction process is always improving and you're more likely to get a better result the next time.

Forecasting – the Foundation

Good forecasting is essential for good decisions. We've looked at sales forecasting, a relatively common example, and you already have some experience from previous years. The same principles apply whether you are launching a new product, buying out a competitor or introducing a new process.

As we've seen, you're making predictions, whether you realise it or not. It's much better to formalize the process and work to improve it. You'll get fewer surprises.

A Balancing Act

Good forecasters balance many factors that are pulling them in opposite directions. They can't be overconfident or underconfident. They can't overreact or underreact. They can't consider too many factors – or too few. They must also be good researchers. They must dig *behind* the news headlines to find the truth.

It takes patience, effort and knowledge to be a good forecaster. It's not easy. If *you* don't have the time or patience to develop the skills yourself, you should create a group of people who can perform the function in your organization. But you must give them the time and resources to get it right.

Feedback Loops

Feedback is a key element of the forecasting process, but it also has wider applications. It *should* be an integral part of every decision-making process. The following diagram shows what happens when you have no feedback:

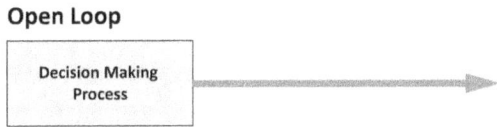

This is called the *open-loop* case. You've made a prediction or decision, but you haven't checked to see what happened. That means there's no feedback to improve the process. Your System 1 assumes everything went well. If there were mistakes, it's *more* likely to repeat them the next time.

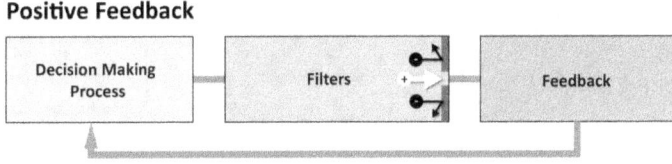

In the second case, you've followed up but you're only looking for *positive* outcomes. This is *confirmation bias* – you've filtered out any negative results. That'll give you a lopsided view of what's happened. It could lead to overconfidence and, consequently, you could make even *worse* decisions in the future.

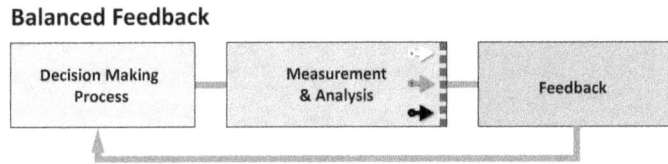

In the final case, you're getting *balanced feedback* – both good and bad. You're measuring the outcome, gathering data and analysing it so that you can be as objective as possible. You're using the feedback to improve your decision-making process, so your next set of decisions will be better. We'll be looking at this again in a later chapter.

Chaos and Predictability

In April 2018, Edward Ott and several collaborators at the University of Maryland claimed they'd harnessed a machine-learning system to predict the future state of a chaotic system at a level of precision never seen before. If this is confirmed *and* found to be significant, it has major implications for business. For example, the stock market and the economy are both chaotic systems. Anyone who can predict stock prices or exchange rates with *any* degree of certainty will have a significant advantage over those who can't. It's worth monitoring progress in this area.

FURTHER READING

Tetlock, Philip, *Superforecasting: The Art and Science of Prediction*, Random House.

10 STUPIDITY

It's time for a quick quiz. Can you answer the following questions?

- How do the sound and pictures reach your TV set?
- What makes your car go?
- Where does emotion come from?

Write down your answers before reading on (Q10-1).

Ok? That was level 1. Now let's go to the next level:

- What kind of modulation is used in your TV?
- What's the primary regulating mechanism in a fuel injector?
- What is the relationship between the amygdala and the hypothalamus?

Write down your answers before reading on (Q10-2).

Still ok? Let's go to level 3:

- What is a concatenated channel coder?
- What is the B.S.F.C. of a naturally aspirated engine?
- What's happens if the hypothalamus detects leptin?

Write down your answers before moving on (Q10-3).

How Did You Do?

You probably answered the first level questions easily enough. You might have written "radio waves" for question 1, "the engine" for question 2 and "the brain" for question 3. But once you start to drill down, something interesting happens. You reach a point where *you don't even understand the questions*. That's when you *should* realize that you only have a *superficial knowledge* of the subject. You don't really *understand* anything about it.

Unfortunately, it's not just one subject or a handful. This applies to *every* subject. That's a very important point and bears repeating: you (and everyone else) have a *shallow* understanding of the world. In a diagram, it would look something like this:

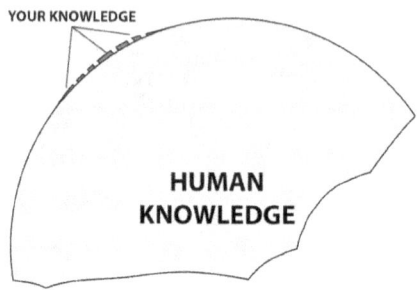

Of course, in reality, the circle would be *much* larger. The dots on top represent your knowledge. They skirt the surface of many subjects, but there's little depth and there are blanks in many areas. We talk about information arriving on radio waves, but we don't really *understand* what that means. When we try to explain something, all we're doing is using *labels*.

Labels

What are *labels*? We've been using them since we were born, and they've proven very useful. We learned the labels "Mum" and "Dad", "table" and "chair" and "dog" and "cat". Once we had a label, we could refer to the object and communicate our desires and feelings about it. However, we usually stopped there. We didn't try to understand the *essence* of these things. We didn't take the TV or the telephone apart to figure out how it works. We were happy to accept the labels – and we still are.

The Problem

Our lack of depth of knowledge isn't usually a problem, but it becomes one when we need to make important decisions. Here's a whimsical story to illustrate. It's 1960 and aliens land on earth. You help them fix their spaceship and in return, they ask if you'd like to accept a gift. They're willing to eliminate rust from the planet. No more corroding bridges, cars or tools. Should you accept or reject the offer?

Write down your answer – and specify how long it took you to reach it – before reading on (Q10-4).

It sounds like a no-brainer – and it is. If you said "yes" it would be a terrible choice. "Rust" is an oxidation process. It applies to iron and other metals, but it *also* applies to metalloids. It gives us silicon dioxide, which is widely used in the semiconductor industry. Without rust, we wouldn't have personal computers, the Internet or smartphones. However, you wouldn't know that in 1960 unless you were one of the few people working on transistor research. There are several important points illustrated by this example:

- You often don't have the information to make a good decision.
- You don't *realise* that you haven't enough information.
- You won't make the *effort* to find the right information.
- You're *willing* to decide without waiting for more information.

Afterwards, you may not realise you've made a terrible mistake. People often run into the *depth of knowledge* problem when they're making a significant purchase. They want to buy a computer but don't understand the difference between processors, or memory and hard disc space. They want to buy a house but don't know what kind of mortgage to get or how to interpret BER information.

When people don't have the knowledge, they often rely on others. That doesn't always work out very well. Keep that in mind, as we explore your *knowledge* in more depth.

Rationalization

Let's address the mental objections you've probably raised when you failed to answer the pop quiz. First, you argued that the missing knowledge wasn't important or relevant. You don't *need* to know about fuel injectors or modulation or amygdalae. You have no interest in these things. You followed up with the argument that if you *did* need that information, you could either look it up, or get someone – an expert – to explain it to you.

Those seem like reasonable objections – but what you've done is "*rationalize*" the situation. You're offering excuses for your lack of knowledge. You've taken the easy route. *You've focused on these few examples rather than the wider issue.*

The key point is your lack of in-depth knowledge on *every* subject.

You're a CEO – a business leader – so it's reasonable to expect that you have an expert level of knowledge in business and management topics. You *should* be able to answer the following:

- In distribution, what is the MODI method?
- What is a fundamental matrix in a Markov process?
- Where is the exponential probability distribution applicable?

> **Write down your answers before reading on (Q10-5).**
> **If you're not sure, just write "I don't know".**

The answers to one or more of those questions could be useful to your organization. Those and other similar techniques could improve performance. If you're not aware of them, it's reasonable to suggest that in an area where it matters, *you have a limited depth of knowledge.* Your lack of knowledge could be causing your organization to lag behind its competitors.

The Knowledge Delusion

I'm sure you're still raising objections. Perhaps you think those things are too complicated for *your* organization. Maybe you don't think you need them or maybe someone else in the organization is looking after them?

Those objections are important because they highlight another of your unconscious beliefs – *if I don't know something it can't be important* – like that pesky Markov process. We *all* believe that we know more than we do. In fact, subconsciously we believe that we know everything we need to know. That's called *the knowledge delusion.*

Answer the following questions:

- What's the best way to tackle the Middle East problem?
- What should the US government do about its deficit?
- How can your team win the championship next season?

Write down your answers before reading on (Q10-6).

These are all complex questions but if you have a general interest in these areas it's likely you also have a solution and it's a relatively simple one. That's the *knowledge delusion* at work. You think you know better than all the experts involved. You might remember that Donald Trump believed the same thing about healthcare legislation!

Knowledge Quality

To recap, *we think we know enough about everything*, even though we don't have in-depth knowledge about anything – even things we're supposed to be expert on.

But what about the knowledge we *actually* possess? Where did it come from and how has it been validated? We gained our early knowledge from our parents, our siblings and later, our friends, peers, teachers and co-workers.

We started with simple *labels*: table, chair, sweets and Mum and Dad. Later we learned about fairies and monsters and Santa Claus. We also learned the difference between good and bad and how things worked – after a fashion. We learned that there was no point asking too many questions because people didn't always have the answers. *We learned that there was an acceptable level of ignorance.*

The foundation of our knowledge was laid down when we were children and it stays with us. That's problematic, because it's the time when we're most impressionable and least critical. Even now, it's hard to accept that much of what we've learned is wrong. Let's take religion, for example. A quick search on Wikipedia brings up the number of adherents to the four most popular beliefs:

- Christianity   2.2 billion   31.50%
- Islam   1.6 billion   22.32%
- Non-religious   1.1 billion   15.35%
- Hinduism   1 billion   13.95%

Those figures mean that no matter who's right, *most of the world is wrong.* For example, if the Christians are right, then nearly seventy percent of the world's population are wrong.

Of course, no matter what your belief, you think *you're* right. But how do you know? Your core religious beliefs were formed when you were a child and are primarily a function of where you were born. If you were born in India, there's an 80% chance you're a Hindu. If you were born in the Czech Republic, there's a 78% chance you have no religion and if you were born in Romania, there's a 99% chance you're a Christian.

Your religious beliefs are the most obvious elements of the belief system that you've been exposed to, but it includes many other factors – for example: what you think of the opposite sex, how you believe ethnic minorities should be treated and whether workers should be trusted.

Valuation

You might argue that any childish beliefs have been displaced by contact and discussion with other people. That *might* be partially true, but how can you be sure? Are you open to other points of view or do you tend to shut them off – to reject them without giving them due consideration?

In the United States, there's been a huge controversy about "fake news", particularly in the last few years. If you were cynical you might say that people define "fake news" as any news they don't want to hear. It appears that Republicans and Democrats only listen to news sources that support their own point of view. One station will proclaim that the president is doing a great job. Another will claim he's a total disaster. How's that possible? The pro-president station will ignore anything that could make their champion look foolish or inept. That incident where he insulted the German chancellor won't be mentioned – instead, they'll declare that the Dow Jones has gone up after his speech.

The anti-president station will look for every slip they can find that indicates he's useless at the job. They'll analyse every utterance to find things to criticize. So, without telling blatant lies, it's easy to create two entirely different views of the same person doing the same job. It just depends on what people are encouraged to focus on – and *we operate just like those stations*. We search out and latch onto any information that confirms our biases. We ignore, reject or trivialise anything that contradicts it.

Business News

Where do you get your *business* knowledge from? Newspaper and magazine business articles are often written by journalists who have tight deadlines and little time to carry out a deep analysis. They depend on the prognosis of industry experts and company spokespeople, many of whom have their own agenda to promote.

To make matters worse, some newspapers also have an agenda. Couple all of that with the reluctance of the few informed people to say too much and you have a recipe for misleading information. So, what makes you believe that *your* sources are dependable?

Fake News

Misleading information pops up in areas as diverse as entertainment, current affairs, science and health. Science writer and author Ben Goldacre has shown how bad information can cost lives. We shouldn't be surprised if someone who gets their information exclusively from newspapers and TV has a warped sense of reality – but we all have biases. We select what to read and watch and hear and we select what we *want* to believe. Then we look for more information to reinforce our beliefs and reject information that might challenge them. We're very good at reinforcing our biases.

Complexity

We even take things a step further because we dislike complexity. *We want to understand everything in simple terms*: black and white, good and bad. Soundbites. What we end up with is *not* the truth. Our simplified model is like a child's motor car. It has the outward appearance of a real automobile, but it lacks the detail, complexity and functionality of the original. It's also a lot smaller!

Ask someone about a work colleague and they might say, "She's very bright but she lacks the confidence to get her points across." That sounds insightful, but the truth is *far* more complicated. Perhaps she has more knowledge than your informant about *one* subject and that's the reason he thinks she's bright. Perhaps she didn't really believe in the argument she was making, on the occasions when she appeared to lack confidence. It might take an entire book to do her justice, but we make do with a single sentence.

Emotional Knowledge

How do you react when you think of the following?

- Socialists
- Communists
- Capitalists
- Religious fundamentalists

These are people working under different economic, social or religious systems. I'm sure you have strong views about some of them and it's likely to link to an emotional response. That makes it hard to separate your emotion from the objective reality. If you believe that a class of people are bad, or fanatics, or mindless idiots, you're bound to have an emotional response and that response is likely to colour your judgement *and* your actions. If you meet one of them, you might find that they're not what you expected – but it's also possible you'll find they're *exactly* what you expected. That's because your prejudices and reactions could bring about your expectations.

Our knowledge is intertwined with emotions. Some things bore us; just the words "bank reconciliation statement" can do it for many people! Some things scare or excite us. And some things are so enticing that we dwell on them for hours. We need to be aware of these emotions and understand what they're doing to our decisions. Please answer the following questions:

- Do you *actively* listen to people before coming to a decision?
- Do you *actively* look for alternative points of view?
- Are you willing to change your mind if someone makes a good point?

**Write down your answers before reading on (Q10-7).
We'll come back to your answers later.**

Knowledge and Your Job

Data, information and knowledge are the raw materials you need for decision making but – as we've seen – there are several problems:

- You think you already know *everything* you need to know.
- You've got no depth of knowledge about anything.
- The *quality* of your knowledge is questionable.
- Your knowledge has been filtered by your prejudices.
- Your knowledge is a vastly simplified version of reality.
- Its collection and use is dictated by your emotions.
- You find excuses to avoid tackling these issues.

Now, the interesting thing is your immediate reaction to that list. Ask yourself these questions:

- Do you believe each of those statements?
- If not, can you check if they're true or not?
- If they *are* true, how are they affecting your decisions?
- Do you think it's worth doing anything about these issues?

Write down your answers before reading on (Q10-8).

How Do You Proceed?

The first thing is to *accept* there's a problem. If you don't, you'll continue to work with knowledge that's not fit for purpose. When you make mistakes, you'll ignore them, forget them or blame others.

If you can accept that there *might* be a problem, you can start using some of the tools outlined in this book. These tools will help you question outmoded beliefs and give you structures to make better decisions. So, let's look at a classic decision-making tool, and then we'll introduce a new one.

The 80:20 Rule

I'm sure you've heard of the 80-20 rule. It was originally formulated by Italian economist Vilfredo Pareto in 1896 and it states that 80% of the effects come from 20% of the causes. He used it to show that 80% of Italy's wealth belonged to 20% of the population. It's been successfully applied to many situations and it can help you.

For example, you might collect data and find that 80% of your profits are coming from 20% of your customers or that 80% of your problems come from 20% of your suppliers. That analysis can strengthen your decision making. You might decide to support your best customers and get rid of those that are dragging down your profits. Now, are the following statements true or false?

- Exercise is good for you.
- Procedures are good for your business.
- Training is good for your organization.

Write down your answers before reading on (Q10-9).

Models

Before we look at the answers to that question, here's another one. You've been given three data points. You've plotted them on a graph and now you want to forecast what's likely to happen in the future. When do you estimate the results will reach 100%?

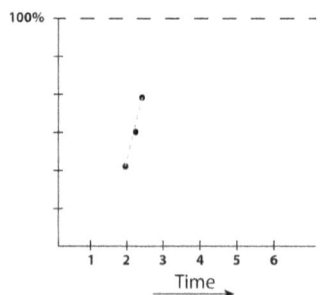

It's natural to look at that graph and assume the result is linear and that it'll reach 100% around time = 3. However, that doesn't have to be the case and, in real life, it often isn't. The graph below uses the same points, but they tell a different story:

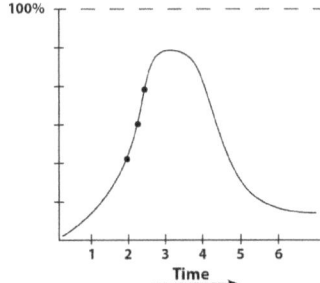

In this case, the results *never* reach 100%. Of course, you couldn't predict the entire curve from those three points, but you *should* have asked: "Are we dealing with a linear situation?" The answer will make a big difference to the forecasts you make.

The first graph represents many people's idealized view of the world. It's simplistic and orderly. All quantities get better or worse in direct proportion. Your System 1 uses this view automatically unless it's explicitly overruled.

The 4-1-1-4 Tool

If you said "true" to any of the three statements earlier (Q10-9), then you also implicitly accepted the assumption "More is better". For example, the more exercise you get, the better it is for you. And that brings us to the 4-1-1-4 tool. Like Pareto analysis, it forces you to take a different look at how things work.

Let's start with that statement about exercise. If you get little or no exercise, then your health is at risk. As you start to exercise, the health benefits increase – just like the three dots in the first graph. Keep on exercising and the benefits continue to increase... but then something new happens. First, the rate of increase starts to slow and eventually, as you continue to increase your exercise, the benefits stop coming and start to fall away. More exercise is making you less healthy!

You can see this on the 4-1-1-4 tool:

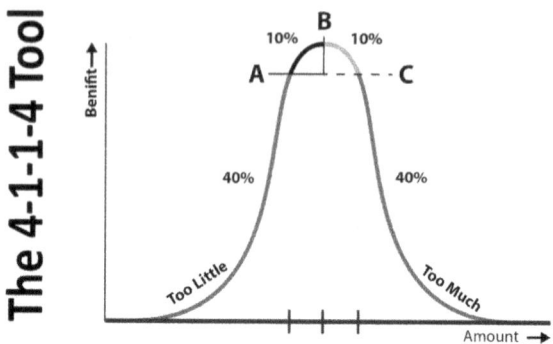

Up to point "A", an increase in exercise led to an increase in benefits. If you can stay between point "A" and "B", then you'll get maximum results for minimum effort. If you increase your exercise between points "B" and "C", you're still getting the benefits but you're putting in extra (unnecessary) effort. After point "C", you're losing out.

The same is true for procedures, for training and for almost anything else you can think of. Our world is *not* linear, and you can use the 4-1-1-4 tool to remind yourself of that. So, the previous statements should read:

- *Some* exercise is good for you.
- *Some* procedures are good for your business.
- *Some* training is good for your organization.

In other words, *more* is rarely better. If you can say "too much X is bad for you" (like "too much food is bad for you") then you've got a non-linear situation. You need to find the sweet spot.

Experts

We've already highlighted some of the limitations of your knowledge, but you might think that doesn't matter too much. You can always call on experts to help you when you need them – we've mentioned them a few times already. The trouble is, experts are subject to the same biases and limitations that affect you, and there are some additional issues.

You can appreciate this by considering the following knowledge diagram – a blown-up version of the previous one.

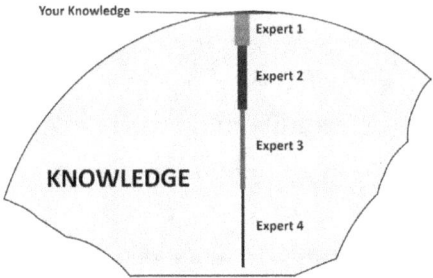

You have superficial knowledge about a subject, but you recognize it, so you call on Expert 1 to help. He's got more knowledge but it's still limited. If you need more knowledge, you need to get Expert 2. But even that's not enough. If you *really* want to understand what's happening, you need Expert 3 and even Expert 4. However, there's a problem – at each successive level, the breadth of knowledge is getting narrower. Expert 4 will be able to tell you a lot about her specialist subject, but she won't be able to discuss related areas.

Evidence suggests that specialist expertise has been overvalued in the last decade or so. These experts are often worse at anticipating and solving difficult issues because they're too narrowly focused. Your challenge is to identify the right experts and get them to work *together* to address your problem.

Functional Stupidity

Sometimes the problem is not just the lack of knowledge but wilfully ignoring its absence. The 2008 financial crisis is a great example of people (including a lot of experts) discounting their lack of knowledge. Banks assembled packages of bad debt mixed with moderate and good debt using derivatives instruments called "collateralized debt obligations" (CDOs). They did this to avoid making huge losses on the "sub-prime" mortgages on their books. Luckily for them, they found people foolish enough to buy them. However, it wasn't long before employees in the *same* banks bought back some of these CDOs at higher prices because they appeared to be doing well in the market! You could consider that as *functional stupidity* on an industrial scale.

If you remember from earlier, functional stupidity occurs when people do their job without considering the broader implications. They don't analysis or question what they're doing. The investors working for those banks saw an opportunity and invested without understanding what the instruments were designed to do. They didn't appreciate the risks they were running.

However, in case you think it's just bankers and investors who suffer from functional stupidity, you should look around your own organization. You won't have to dig too far to find it deeply entrenched. We're going to look at three examples, but you'll be able to find many more.

The Hockey-Stick Effect

You're the operations manager of a manufacturing company. It's coming to the end of the quarter and you haven't met your target. You need to do something, so you bring forward some of the orders from next month to fill the gap. Production increases but it's still not going to be enough, so you authorize overtime and get people to work extra shifts. Eventfully, you get your numbers (usually on the last day) and the quarter is closed off. The production numbers for the quarter look like this:

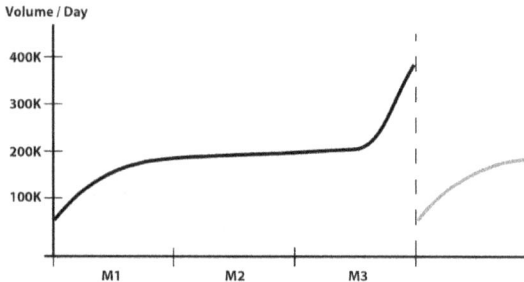

The graph shows that daily throughput was low for the first week then it levelled off at around 200K. However, in the last week of the quarter, it spiked to almost 400K. Now let's consider the story behind the curve: there's little to do in the first week of the quarter so people and equipment are underutilized – those are wasted resources. It takes some time to get to normal production rates and then things settle down.

However, when the shortfall is noticed, orders are brought forward from the next quarter. This causes problems for materials planning and logistics. People are also brought in on overtime and extra shifts are scheduled, so machines run longer and consume more energy. A huge amount of money is wasted on the exercise – all to meet an artificial target. And then, of course, the cycle repeats itself the following quarter – that's functional stupidity.

This example illustrates again how *metrics* can drive dysfunctional behaviour. The production manager is held accountable for meeting throughput targets and he achieves them, but at the expense of company profitability. That shows a lack of understanding and accountability from board level, through the CEO, to the person making the stupid decisions.

Training

We'll cover training in detail in a later chapter. It's enough to point out here that *most training is a waste of time and money*. This qualifies as functional stupidity because the core reasons why training programmes don't work were identified over a century ago!

Reports

Companies generate hundreds of reports but most of them aren't used. Many are generated by computer and just waste processing time, but some are still created by people who must collect and collate the data and laboriously produce reports, sometimes with different versions for departments and managers. This is a total waste of company resources.

Promoting Stupidity

- "The Dow gained 150 points due to the president's speech."
- "Sales have fallen due to lower consumer sentiment."
- "We didn't meet our production numbers because the machines were running badly."

What's your response when you come across statements like these? Do you accept them without question? If so, you've been conditioned to respond with passive acceptance and *you're probably promoting a culture of stupidity in your organization.* Let's take another look at these statements and find a more suitable response:

"The Dow gained 150 points due to the president's speech."

"How do you *know* that? Could anything else have caused it? What about a change in oil prices, a big product release or a thousand other factors? Is every change in the Dow traceable to a presidential statement? What about times he said something positive and it fell, or vice versa? It's likely that causality is being assumed just because two events have taken place in proximity to each other. If so, it's just lazy reporting."

"Sales have fallen due to lower consumer sentiment."

"Yes, our sales have fallen and so has the consumer sentiment index. However, *co-correlation doesn't imply causation.* In other words, just because two things happen at the same time doesn't mean that one caused the other. Look at the figures from last July. Consumer sentiment was low, but our sales were high, and the reverse happened in November."

"We didn't meet our production numbers because the machines were running badly."

"Have you *checked* the machine downtime figures? You can see they're no worse than the previous three months, and you made your target then. Let's look for an alternative reason for this situation."

If you simply accept statements without question, managers will learn to avoid the hard work of gathering data, analysing it and identifying the true situation. They'll just give you the first reason that occurs to them – that's a System 1 response. They'll also pick up the habit and won't bother to question the team leaders. Soon, team leaders will also be accepting top-of-the-head answers from workers and the quality of information will deteriorate. If that's been happening in your organization, there *is* a solution: use insightful questions to prevent the spread of stupidity.

Knowledge Maps

Do you know where key knowledge is contained in your organization? If you don't, you could find yourself in the same position as a small company in southern France. They'd been expanding their business and things were going well. Then one of their senior engineers left the company and, almost immediately, customers started to complain that the quality of the product was unacceptable. The engineering group struggled but never managed to restore it and the company eventually went into liquidation. The key process knowledge was in the head of *one* engineer and he took it with him.

The solution is to construct a knowledge map and use it to identify the critical knowledge in your organization. You can do this using the following hierarchy:

- Area
- Process
- Task
- Knowledge required
- Knowledge contained

Break the organization up into different functional areas, e.g. production, purchasing, shipping, etc. Then identify the processes in each area. In the production area, you might have "component placement" and "solder reflow". Next, identify the tasks required for each process. That might include "machine start up" and "pot maintenance" for the reflow process.

Now identify the knowledge required. There may be two types of knowledge: generic and specific. "*Generic*" is knowledge that can be used in many situations. For example, an electrician has technical knowledge they can use in any company. "*Specific*" refers to knowledge that applies only to a particular process. Finally, for each element of knowledge, where is it contained? Is it written down or does it only exist in the head of an employee?

Once you've completed the knowledge map, you can use it to identify areas of critical knowledge. Then you can put systems in place to capture and control it. That ensures you don't lose irreplaceable information that could impact the organization's competitiveness and survival.

The Experience Myth

You've advertised for an *experienced* long-distance lorry driver and the applicant tells you that he's driven a truck three times. You're looking for an *experienced* plumber and the person tells you she's fixed four leaks. Would you hire those people? Of course not. You'd realise that they are *novices*. However, if a CEO has been involved in the takeover of three or four companies you'd regard him as an expert! There's obviously a double-standard somewhere.

In the first two cases, if the person isn't doing a good job, the results will be immediately obvious. You'll get a crashed truck, product scattered along the motorway or a room rapidly filling with water. After a company takeover, the results might not be known for months or years – so there's little feedback. And there are other differences; the environment for a truck-driver or a plumber is well defined. There are rules to follow and much repetition, so learning occurs each time they do the job. That's not *as* true in complex situations like a business takeover.

While the novice truck driver and plumber are learning, they *must* use their System 2s until they're competent. Eventually, their System 1s will identify patterns and take over the work, but that takes time and repeated exposure. How long before our truck driver can reasonably claim to be experienced? You could argue that it'll take several years, but let's be optimistic and say three months. Assuming a five-day week, that's 60 occurrences. It follows that a manager would need *at least* that level of exposure before claiming to be competent in a *more* complex area. She'd need to be involved in 60 takeovers before claiming to have expertise. Most CEOs don't have that level of experience, so they're still novices. Despite that, they still use their System 1 "gut" to make key decisions. It's not surprising there are many failures.

Real Experience

People often judge others by their *experience*; it's taken as a proxy for *knowledge*. However, as we've seen, just being *exposed* to a situation doesn't give anyone the knowledge or insight to make good decisions. That takes a sustained effort from System 2. A manager could claim to have 12 years' experience but that might mean they've learned just enough to get by in the first year – and lived on that knowledge for the following 11 years! The amount of experience someone has, measured in years, means very little. It's the expertise they've built up that's important – and you should be able to *test* for that.

FURTHER READING

Goldacre, Ben, *I Think You'll Find It's a Bit More Complicated Than That*, Fourth Estate.
Alvesson, Mats & Spicer, Andre, *The Stupidity Paradox, The Power and Pitfalls of Functional Stupidity at Work*, Profile Books Limited.

11 DECISIONS

We're going to look at several ways to improve your decision-making, but first, let's see what you've already revealed about yourself. At the start of Chapter 3 (Q3-1), you were asked if you made big decisions easily. Did you answer with a high number (over 60%)? You'd think that would be desirable – but it isn't. If you're making significant decisions without too much effort, you're also making mistakes. The reason is simple – you're using System 1 instead of System 2. You're going with your gut. System 1 is fine for some decisions and some professions (like drivers and firefighters) but it's not appropriate for the decisions *you* should be taking. Remember, System 1 is only useful for making re-occurring decisions in a controlled environment and only when there's rapid feedback. If that describes most of the decisions you're making, then you're making the wrong ones.

Decision Classes

How many business decisions do you make every day? If your answer is more than three or four, then you need to reconsider your role because you're too involved in the day-to-day operations. You aren't delegating enough. One way to improve your decision-making process is by examining all the decisions you're currently making and transferring as many as possible to other people.

It's always a good idea to push the level of decision-making down the organization. If an operator can make a decision, then he doesn't have to wait for a team leader to make it. If the team leader can make it, she doesn't have to wait for a manager. That results in a more flexible, responsive and effective organization. You'll find it difficult to let go of many decisions, but you can provide written policies for team leaders and managers so they understand the criteria to use when making them.

However, be warned: even with guidelines in place you'll find that people will make mistakes. That's a normal part of learning, so it's important not to overreact. You'll have to tolerate mistakes for a while until people get used to their new responsibilities. It will eventually free your time for important decisions.

Key Decisions

Let's consider the type of decision *you* will be making. You've been worried about losing market share because Jones's prices are lower than yours. So you came in on Monday morning and announced to your managers that you're going to buy Jones's business and combine product lines. The managers agree that it sounds like a great idea. What have you just done? Here's a quick summary:

- You've fixated on one problem
  (your competition is undercutting you)
- You've fixated on one solution (take them over)
- You've reduced the chance of getting good advice
  (by announcing your decision)
- You've psychologically tied yourself to the solution
  (you'd be embarrassed to change your mind)

CEOs frequently make decisions this way and the consequences can be serious. Here are some of the tasks that'll result directly from your weekend brainwave:

- Set up a team to analyse Jones's performance
- Engage lawyers to work on anti-competition issues
- Set up an IT team to identify system integration issues
- Set up a marketing team to determine how to promote the combined brands
- Set up a negotiation team and develop a negotiation strategy.
- Identify areas of overlap in production capability

Your single decision could result in thousands of hours of work. When external resources are used (like the lawyers), there may be substantial costs. When internal resources are used, there could be a negative impact on existing operations, and improvement projects could be sidelined. There are also hundreds of additional decisions that must be taken that'll soak up management time. If the takeover goes ahead, there could be more significant issues. Many companies have seen their stock price collapse after an unwise takeover.

Of course, the decision doesn't have to be about takeovers; it could be about creating a new product, developing a new marketing strategy, buying new machines or anything else that could have long-term consequences. So, how should a CEO approach those decisions?

The Trigger

There's usually a trigger for a decision. It might be a problem, an opportunity or a deadline. In our example, the problem was the perceived danger to your market share by Jones's price undercutting. The first question to ask is this:

Is This Really a Problem/Opportunity/Deadline?

In other words, will the price that Jones is offering really affect your market share? Perhaps they're selling at a loss, in which case you might want them to continue until they drive themselves out of business. Even if they're making a small profit, you might find that it's not worth the effort to chase that section of the market. Perhaps they're not your core customers? And even if they are taking market share, is it worth spending millions to get it back? Does it justify that investment? It might take some time and effort to get answers to these questions, but the commitment is small compared to the alternative. It may turn out there's no need to worry about Jones because they're not really a threat.

Alternatives

Next, let's assume that the problem, opportunity or deadline is real. You need to act. Authors Dan and Chip Heath point out that one of the most dangerous questions you can ask yourself is the "either/or" one: should I do it or not? "Should I take over Jones's business?" focuses you on a single decision and blinds you to everything else. There are many other options available, but you won't see them unless you widen your perspective. So, take an alternative approach: if it's going to cost €10 million to buy the company and another €2 million in costs and fees, ask yourself, "What else could I do with €12 million?" For example, could you:

- Buy automated machines to reduce cost?
- Conduct a series of marketing campaigns?
- Develop a new technology?
- Create a new product?
- Build a new plant in a low-cost country?

There are many other alternatives and some of them are likely to be far superior to your initial choice, but they won't spring instantly to mind. You'll need to actively look for them. So, how do you do that reliably? Here are the first seven steps for your new decision-making process:

1. Get a team of people involved
2. Explain the trigger in generic terms
3. Ask each person to come up with 10 suggestions
4. Ask them to come back with a list
5. Get them to create selection criteria
6. Review and score the suggestions
7. Select three options for detailed analysis

1. Involve Other People

The first step is to set up a working group with five or six members. It's important to involve other people as early as possible in the decision-making process, because each will have a different perspective. That'll widen the range of options you'll get to consider. Some suggestions may be predictable but there may also be many unexpected suggestions, because people have different backgrounds and interests.

2. Explain the Trigger

You need to explain what you want people to do, but not to lead them to a specific solution. You can do this by explaining the trigger in factual and generic terms. Consider the following statement:

"Our sales have dropped by 5% over the last three quarters. Our competitor's sales have increased over the same period. Their prices are 10% lower than ours. I'd like you to come up with some options to tackle this situation."

Is that a good way to raise the issue? It's factual and there's no mention of Jones or a takeover, so that's positive. However, you must be careful about the facts you select. For example, did you notice that there's an implication in the statement? It assumes that the reason for the drop in sales is the price difference. That may not be true. It could be due to the quality of the competitors' products or their clever marketing strategy. If you're not absolutely certain about the significance of a fact, like the impact of that price difference, you should leave it out.

3. Get Suggestions

One of the major issues that affects the quality of the decision-making process is "*sufficiency*". That means giving up too easily. If you were asked to come up with a solution or an innovative idea you might offer two or three suggestions. When pressed you might be able to offer two or three more. After that, you'll begin to find it more difficult. The reason is that your System 1 doesn't have any more matches available. That means your System 2 needs to get involved but, as we've seen, it's lazy. It'll try to convince you that the suggestions you've already come up with are sufficient and that one of them is the best possible answer. The problem is that those suggestions are the obvious ones. If you select one, you'll be settling for second (or twenty-second) best. To avoid that happening, you need to jump-start System 2 into action. One way to do that is to force it to come up with more suggestions.

That's why you'll ask each member of the team to come up with 10 suggestions for the next meeting. Some people will react negatively to that request, even after you explain your reasons. They'll find excuses for not getting involved. You must insist on their co-operation or remove them from the team (the last resort). To make it easier for them, you can tell them they can get help from people outside the meeting.

4. Come Back With a List

Once everyone understands what's involved, you should ask them to come back with a list of suggestions for the next meeting – which should be scheduled within two days. That'll give people enough time to think about the issues, talk to other people and come up with suggestions. If you give them more time, they'll tend to put off doing anything about it until right before the meeting! Each list must be submitted in an identical format and handed to a nominated person on the team who will collate all the suggestions. The lists are submitted anonymously, and that allows *you* to submit suggestions if you choose. The group will now discuss the suggestions and may also come up with new ones, or combine existing ones.

One of the reasons for approaching decision-making in this way is to prevent "groupthink". This is where a group of people come up with a single solution without seriously considering alternatives. When that happens, they'll overestimate the chances of success, ignore or marginalize anyone who might disagree with them and create an artificial group consensus. None of the participants will want to rock the boat by questioning the group's decisions. They'll ignore potential problems and become blind to any negative consequences. The result can be catastrophic. A famous example of this phenomenon was the CIA-sponsored invasion of Cuba at the Bay of Pigs. President John F. Kennedy and his inner circle ignored obvious problems, including the failure to destroy the Cuban Air Force. They allowed the invasion to take place with disastrous consequences.

5. Develop Selection Criteria

The next step is to select the most promising suggestions, so they can be considered in more detail. However, there are two important stumbling blocks:

- Over-choice
- System 1's selection bias

Over-Choice

Humans don't cope well with too much choice. Research has shown that when we're faced with many options, we often take the easy way out and refuse to make *any* choice. We revert to the status quo. In our example, you've deliberately generated a large number of options, so how can you make sure your team aren't overwhelmed? The answer is to carry out a systematic pruning operation.

Selection Bias

The second problem is that your System 1 will want to select a familiar-looking solution. Something that'll fit in with what it already knows. When faced with a range of options, it's likely to select the most familiar, not the best. This is an example of its tendency to substitute an easy question for a difficult one and answer that instead. The difficult question is: "Which of the alternatives is the best?" The easy question is: "Which of the alternatives am I most comfortable with?" You must take care to avoid answering the wrong question.

The Decision Matrix

The best way to tackle these problems is to use a decision matrix. Start by asking the team to assemble a list of criteria that's deemed important with appropriate scores. For example, if you were buying a car, you might select the following:

- Price
- Exterior colour
- Interior colour
- Engine type
- Engine size
- Shape
- Acceleration
- Fuel consumption

Some of these will lead to an immediate pruning decision. For example, if the price of one model is €80,000 and that's over your upper limit, you can eliminate it and it needn't be considered further. If some criteria are more important than others, you can apply a weight to each of the criteria. For example, if acceleration is more important than fuel consumption, you might apply a weight of 5 to fuel consumption and 8 to acceleration.

You should try to include *all* relevant criteria and not overlook anything important. For example, the list above omits an important factor for many people: *prestige*. Even if you think you're not affected by emotional factors, you should include any relevant ones with a low rating. A typical decision matrix will look like this:

Criteria	Weight	Model A	Score	Weighted Score	Model B	Score	Weighted Score	Delta
Price (Max. €60,000)	8	€54,000	6	48	€57,000	5	40	8
Acceleration 0-60	8	10.5 sec	6	48	13.5 sec	3	24	24
Engine type	7	Petrol	7	49	Diesel	5	35	14
Engine size	7	1800cc	7	49	2000cc	7	49	0
Fuel consumption Mpg	5	33.5	4	20	45.3	6	30	-10
Shape	4	Coupe	8	32	Saloon	6	24	8
Exterior colour	3	Black	5	15	White	4	12	3
Interior colour	2	Grey	3	6	White	4	8	-2
Prestige	2	Average	5	10	Average	5	10	0
		Total		277			232	

Selecting the Criteria

So how do you choose the selection criteria? You ask each person on the team to write down the ten most important factors and rate them. They should consider both positive and negative factors. For example:

- Throughput increase (Positive)
- Quality increase (Positive)
- Ease of use (Positive)
- Cost (Negative)
- Disruption (Negative)
- Risk of failure (Negative)

They must do this *before* any discussion, so they're not influenced by yourself or others in the group. The lists are then collected from each person, the results are combined, and the team can then start to consider all the suggestions. *It's very important that you should not express a preference or reject anything.*

Now, for each criterion, have a team member explain why it's important. Allow a short discussion to take place and then ask everyone to rate it from 1 to 9. Then average the results for each person and you will end up with a list like the one in the left box of the table above. But hold on, what about your input? This is the difficult part – *you don't have any*! The reason is simple. If you express a preference or a direction, your managers are likely to be influenced by it. They won't waste time looking seriously at other options if they think you've already made up your mind. So, the question you must ask yourself is: "Which of the following options would I prefer?"

- To make the decision myself?
- To get the right decision?

Yes, the irony is not lost – this *is* an either/or decision, but research shows that the answers are almost always mutually exclusive. *Your prime responsibility is making sure the process works.* You must ensure that no single person has too much influence and everyone has contributed properly. You must prevent groupthink. If you can do that, you have a much better chance of avoiding the biases that lead to mistakes. In other words, *you're more likely to make the right decision if you give up the right to make it yourself.*

6. Review and Score the Suggestions

Now, go back to the list of suggestions and put them into the matrix. Then get the team to discuss them and score each one. If there's disagreement about any item, try to resolve it, but if that's not possible, you should use the average score from the team. You'll end up with a total score for each option. All you need to do now is eliminate the options with low scores and you're left with the top three options.

7. Select Three Options for Detailed Analysis

The decision matrix allows you to consider many options quickly. However, you might find that your preferred option didn't make it to the top three. You might then be tempted to "fine-tune the criteria" to include it or use "executive privilege" to override the matrix completely. Both are bad decisions. Can you see why?

Both decisions would undermine the credibility of the process. They'll send a very clear signal that you're not serious about it. The next time, your team won't take it seriously either.

Research has shown that a good algorithm will usually outperform human judgement. In other words, it's far more likely that the result from the decision matrix is correct.

You must trust the result of the decision matrix. If the criteria that the team selected and the weighting factors are correct, then the result should also be correct. That's why it's important to take enough time to select the right criteria and apply the weights carefully. The advantage of this process is that it forces you and your team to consider all the criteria and a wide range of options.

Emotional Reaction

Are you feeling constrained by this approach? Are you ready to dismiss it? You may feel that using a matrix to actually *make* the decision is going too far. *That's your emotion talking.* The matrix is a tool. It's designed to take emotion out of the decision-making process and help you make System 2 decisions. If you have a problem, it's because your System 1 has a strong hold. You prefer to *react* rather than make logical decisions. You prefer to rely on your gut. It's normal to resist using tools like this. As humans, we've evolved to react quickly to situations. A rustle in the grass and our ancestors were off running – there was no "thought" involved. Those who stood around thinking didn't survive long enough to pass on their genes. We're the descendants of fast reacters, so we're genetically programmed to make fast "gut-reaction" decisions. Of course, the environment has changed but our instinctive decision-making process (System 1) is still in control. There are still many threats but they're far subtler than a sabre-toothed tiger and your instinctive reaction could be exactly the wrong thing to do. You must decide if you want to stick to a pre-historic decision-making system or use something a bit more modern!

Bad Decisions

In case you think that having a logical decision-making process isn't necessary, just consider how many bad business decisions are being made every day. A 2004 study showed that *83% of mergers and acquisitions fail to create value for the shareholders.* Not all of those were *total* failures but they didn't repay the enormous effort that was put in. There's an opportunity cost in all these cases. Another study found that 60% of executives believe that bad decisions are as frequent as good ones in their company. But perhaps you don't remember any bad decision being made in the last year or two? Just ask your managers and staff. They're far more likely to remember your bad decisions for two reasons: first, they didn't make them, and second, they must live with the results. Don't forget, your brain protects you from negative outcomes though selective memory loss. You're more likely to forget the bad results.

Next Step

Your team have considered many possible options and narrowed the results down to three using the decision matrix. Now, they need to dig deeper. They need to collect more data, identify and test assumptions and, if possible, run some small-scale tests to see if their assumptions are correct. In other words, do what they can to remove subjectivity from the decision. They'll continue to fill up the decision matrix, adding new rows for each new result they uncover.

Confirmation Bias

That all sounds terrific, but you may recall the problem we have with *confirmation bias.* This is where we only look for data and information to confirm what we already believe. How can you prevent that happening?

One way is to nominate a member of the team to act as *devil's advocate.* He must argue that it's a bad idea and collect data to prove his case. He should only use facts and probabilities based on the available data. Of course, it's important that *all* the data he finds (including any positive results) are made available to the team.

Facts and Probabilities

The final decision should be based as closely as possible on facts and probabilities and as little as possible on emotions. For example, if you're looking at takeover options, you might look at future sales. You should estimate what's likely to happen by looking at the probabilities:

- Sales stay flat. Probability: 50%
- Sales drop by at least 20%. Probability: 35%
- Sales increase by at least 20%. Probability: 15%

How do you get those probabilities? You can't do it by just looking at your own company. Instead, you must get an *outside view*. Get your staff to research what's happened to sales in similar companies over the last five years. That'll give you a base rate that you can work with. You should also use all the techniques outlined in the forecasting chapter to make sure that any predictions you make have a solid basis.

Final Review

Once you have the facts and the probabilities, you're ready for the decision-making process. First, confirm with the group that the three options are still the best possible. Make sure that nothing has been uncovered to suggest that another option should be considered.

Now, carry out a final review of the selection criteria to make sure they're all still valid. Pay attention to those that have been added since the first review. Also, see if any criteria are being overlooked. Could there be other factors you should consider before making the final decision? For example:

- *Stakeholders*: Who will be affected by the decision?
- *Costs*: Have you considered direct and indirect costs?
- *Timeframe*: Is your timeframe realistic? What would happen if it took twice as long to implement?
- *Resources*: Have you underestimated the resources required?

Assumptions

List all assumptions you're making about the *trigger* and the solution. For example, the trigger in our takeover scenario was a drop in sales.

We *assumed* it was due to Jones's low prices, but has this been proven or is it still an assumption? We *assumed* that a takeover was necessary to remove the threat. Is that still an assumption? If you're left with any significant assumptions, it means you haven't done enough research.

Go for It!

This is where bad decisions are born. You could call it the "*Aw, to hell with it!*" syndrome. People get impatient. The process is taking longer than expected and they think they know the answer anyway, so they just rush ahead and decide with too little data. If you allow shortcuts you're likely to suffer in the long run. It's your job to manage expectations in the group.

Making the Decision

When all the data has been collected and discussed, each person in the group gives a score for each criterion for each option. They do this privately, so they can't influence each other. You collect those scores and then discuss the results. The first aim is to identify any criteria where scores are widely scattered. If some people have scored 2s and 3s while others have scored 8s and 9s, it's useful to understand *why* the discrepancy exists. People get a chance to change their score – but only if they're convinced by the opposing argument. The intention is to get clarity and score accordingly – not to get concessions.

Once the discussion has been completed, the scores are added up and averaged for each criterion for each option. The weighted score for each criterion is then calculated. Finally, these are added up and the results are shown to the group. The decision has been made.

Your New Role

Once again, it's important to emphasise that your role is *to make sure that the process is effective, not to make the decisions*. You're trying to prevent your System 1 from coming to a snap decision based on a misguided belief in its own invincibility. You're also trying to prevent everyone else's System 1 getting in the way. You're the only one who can do that, because you're the one with the authority.

Now, you must make sure that everyone in the group fully understands what the decision entails – and find out where they stand. That'll take more discussion. The ideal situation is that everyone is fully behind it, but at the very least, you need to be sure that no one is going to undermine it. So, don't express any opinion yourself but get everyone else to state their opinion. If there are objections, try to establish the reason why. Some may be logical but there may also be hidden reasons – for example, a perceived loss of prestige or power. You'll need to address those reasons, particularly the hidden ones, but you may need to do it on a one-to-one basis. Then we come to the final step in the process.

The Red Team

Select five or six people who have *not* been involved in the decision-making process. Tell them that you want them to help evaluate a decision. You'll then do the following:

- Select a member of the original team to explain the trigger
- Select a second member to list the three options
- Select a third to *explain* the decision
- Have the remaining members explain the *reasons* for selection

The red team can now ask clarification questions. When they're finished, they should go away for two hours to consider the decision and its ramifications. At the end of the first hour, they have the option of coming back to ask additional clarification questions.

At the end of the second hour, they come back and face the original team. Each red team member can ask five questions each, using the form, "Have you considered…" For example, they might ask, "Have you considered what will happen if the other competitors drop their prices when you take over Jones?"

The original team members can answer "yes" or "no", but no further discussion is allowed. All questions are recorded. At the end of the session, the Red Team members are thanked for their participation, and leave. The original team now go through the questions again – this time ticking off those that have been properly considered in the matrix. If a significant question has *not* been properly addressed, that's taken as a red flag. It means that the original process was not thorough, so *everything* needs to be evaluated again – not just the item that was identified. If one item can be overlooked, so can others. If nothing was overlooked, then the decision has been made. You're ready for planning and implementation.

Assumptions

We worked our way through the decision-making process without looking closely at some key factors. Now we need to circle back and consider them. The following are important:

- How do you select the team members?
- What resources should they have?
- How do you train them?
- When should you schedule the meetings?
- What should you record?
- How involved should you get?

Selecting the Team Members

You need a mix of people on the team, people who are willing to stand up for what they believe and to fight their corner. You need people who have an open mind and are willing to learn. And, you need people who're willing to work hard.

What you *don't* want are the following:

- People who always agree with you
- People who want to dominate every meeting
- People who aren't willing to change their minds
- People who have a negative attitude
- People who think they know everything
- People who won't contribute

There should be a mix of expertise and knowledge levels. If appropriate for the decision, you might have operators, engineers, team leaders and managers in the group.

Resources

The most important resource for the members of the team is *time*. You must ringfence specific times when they can work *exclusively* on the decision process (for example, every Tuesday and Thursday). That means someone else must take over their workload. This is a key requirement and if it's not achieved, the process is likely to fail. They may also need someone to help them with research. Members of the team may not be aware of internal and external resources that are available to them. If the decision relates to a technical issue, it may be necessary to train team members on the basic principles involved.

Training

All members of the team must be trained on the principles of decision-making. That should include the following topics:

- Human biases
- System 1 and System 2
- The decision matrix
- Evidence-based decision making
- Simple probability theory

Three or four one-hour sessions should be enough to bring people to an acceptable level. The key point is that it should be *you* who carries out the training. There are three reasons for this:

- Your presence will emphasise its importance
- You're likely to identify anyone who has reservations
- You'll learn more yourself!

The Best Time

It's important to pick the right time for any meeting where your group is going to make significant decisions. Researchers have found that we have a limited amount of willpower available and it depletes through the day – particularly when you're using System 2. This means you should *not* schedule meetings that make significant decisions in the afternoon or evening, because people are more likely to agree to a bad decision. Likewise – you should avoid scheduling important meetings just before lunch, as this has also been shown to be a bad time. To complicate things further, some people work well in the morning, but others are slow starters. Combining results from several studies it appears that the best time for important meetings is between 9.00am and 12.30pm (assuming an 8.00am start and 1.00pm lunch).

Recording

You need to keep a record of the outcome of each stage of the decision-making process. The criteria and options will be recorded on the decision matrix, but you should also track any actions that need to be completed, any questions that need to be answered, and any difficulties or opportunities the group has encountered. You also need to record the reasons for each of your selections. The notes should be short. Make a dedicated person responsible for recording them. That person must be willing to ask for clarification if they don't understand something – and that can be useful for the entire team because there may be misunderstandings about what terms mean. There are two reasons for recording what's going on. The first is to ensure that the team works efficiently together. The notes provide a common understanding and ensure nothing gets overlooked. The second is to continue to improve the decision-making process.

Your Involvement

You're responsible for the decision-making process. You pick the people who'll be the team members, you'll train them yourself, you'll supervise the development of the decision matrix and attend the meetings to ensure that the process is being followed. However, you *won't* pick the criteria or select all the options. You may ask questions, but not in a leading manner. Nothing you do must indicate which criteria or options you favour and which you regard as unimportant. That could unduly influence the team and that would bring back all the limitations of a single decision-maker. In the same way, you must never overrule the final decision. If you do, the entire process will be compromised, and you'll never get the same level of buy-in again.

Dilemma

Right now, you're facing a dilemma. Your System 1 is telling you to reject this approach. It *likes* making decisions. It can't admit that it's making mistakes and hiding bad results. It's also dragging in *emotions*. It's telling you that you won't get *all* the credit if other people are involved in each decision. It's very seductive and it's coaxing you to take the easy option. Do you have the strength to resist? Remember, this can be *your* defining moment.

Suit You?

So, will you use the logical decision-making process or not?

Write down your answer before reading on (Q11-1).

You've been hired to get results – not make decisions. Decisions *must* be made – and they must *right* – that's your responsibility. But you don't have to *make* them. So, let's do a quick reality check. Which of the following options would you choose?

- Make the decision yourself – with a 35% probability of success
- Facilitate other people to make the decision – with a 65% probability of success

The answer *should* be obvious, but some people have mental models that won't *allow* them to select the second option. They can't be World Class leaders because they'll always find ways to hide or justify mistakes. The most common argument they'll make is: "I'm different – this doesn't apply to me".

Different, Really?

Everyone believes they're different – and they're right. We're all different, but, as you've seen, we all suffer from the same biases. We don't realise or accept that we have them, yet they seriously affect our decisions. Some researchers believe it's *impossible* for a single human to make a truly unbiased decision. So, is your System 1 still trying to persuade you to make complex decisions using your gut? Can you see why that could *never* be a good idea?

Problems and Solutions

The decision-making process outlined in this chapter has been created to tackle common cognitive problems. The potential problems it addresses include the following (any one of which is enough to derail your decision):

- Insufficient knowledge
- Insufficient understanding of the issue
- A distorted view of the issue
- Overlooking or dismissing key elements of the problem
- A narrow focus on a single solution
- A solution presented in "yes/no" terms
- A simplistic solution
- An emotional attachment to one solution
- Overlooking or dismissing problems with the solution
- Overlooking implementation issues

The process has features to tackle those problems:

- Get multiple people involved
  (different knowledge and viewpoints)
- Give them a structure (to avoid overlooking anything)
- Get them to confirm that the trigger is valid
- Get them to generate many options
  (to avoid fixating on one solution)
- Get them to examine the pros and cons of each option
- Get them to score options
  (to remove as much subjectivity as possible)
- Get them to dig deeper into the most promising options
  (to uncover hidden issues)
- Ensure all discussions are based on facts and probabilities
  (reduce emotional attachments)
- Use a written history (to make everything explicit)
- Allow the people who will be responsible for the
  implementation to make the decisions

Those are the reasons that this method is better than gut-feel. It *forces* people to use their System 2 capabilities. It's the closest thing to logical decision-making that humans can achieve.

Cascading Decisions

All the issues that make it difficult for *you* to make better decisions also apply to your managers. That means you need to get *them* to use the structured decision-making process in their own areas. For example, when the marketing manager wants to create a new marketing campaign, she must set up a team and follow the decision-making steps. She can't pick the criteria or vote on the options – the team members will do that. Her job is to ensure that the decision-making process is as effective as possible.

The same happens at the next level. When a team leader wants to create a new overtime schedule, he gets his staff involved and they go through the decision-making process. That means you're more likely to get the best decisions at every level in the organization.

No Trigger?

We've been looking at decisions that have been prompted by a trigger but what happens if there's no trigger? Let's say things are going well. There are no major problems and everyone is reasonably satisfied. It looks like there's no need to waste time invoking the decision-making process, right? Wrong! It's very easy to be satisfied with the status quo, but that can be a huge mistake. Let's say you face the following situation:

- Make no decision
- Make a decision with a 55% chance of success

If you're risk-adverse, you might avoid the second option because there's a 45% chance it will fail and there's a cost associated with that failure. Of course, we haven't spelt out what the costs and benefits are yet – so let's do that now:

- If the decision is right, you'll add 20% to the organization's value
- If the decision is wrong, you'll subtract 15% from it

It still looks risky, doesn't it? If you had to decide right now, what would you say? By the way, make a note of any missing information you might need.

> **Write down your answer before reading on (Q11-2).**

There's one more bit of information you didn't have, and most CEOs often don't consider. It's the risk associated with *not* making a decision. Suppose you have discovered that the following probabilities apply if you *don't* make any decision:

- Things continue as they are: 20%
- 20% added to the value of the organization: 10%
- 20% subtracted from it: 70%

Those figures show that you're very likely to lose out if you don't take action. Without that knowledge, your natural inclination could take you in the wrong direction. It's very easy to get caught up in the day-to-day trivia and ignore the big picture until it's too late. System 1 loves repetition and that's why the status quo is often the most dangerous option.

Tripwires

If you've *really* established that there's no need to do anything *now*, you should still avoid getting caught unawares, by setting up "*tripwires*". These are checks that'll warn you when you *must* act. Some are conditional. For example:

- If sales fall below €400K in any quarter
- If a project is more than two weeks behind schedule
- If overtime is greater than 10% of payroll costs

You can also have time-based tripwires:

- At the end of each quarter
- On the first week of the new year
- One month after each project is completed

You can decide what you'll do if a tripwire is "sprung" ahead of time. For example: "If sales fall below €400K in any quarter, we'll advertise our products in the local press." "If overtime is greater than 10%, we'll cut back on tier 2 support." The time-based tripwires are used to make sure that things continue to work well. For example: "We'll continue with this strategy until 15th June. If we're still meeting our targets, we'll let it continue, otherwise we'll set up a review group to develop a new strategy." You can use tripwires to prevent being blindsided by new developments: "We'll review all competitive products in our area at the start of each quarter. We'll carry out a full evaluation of those that have increased market share since the previous review."

One person should be responsible for all the tripwires. Once a day, they'll check to see if any of the conditions have been met, and if any *have* been triggered, they'll inform the appropriate people. It might be possible to automate some tripwires, but it's also important to have a dependable human in the loop so that nothing is overlooked or forgotten.

The Best Leaders

World Class CEOs – the best of the best – have *intellectual humility*. They appreciate the limitation of their knowledge and understand their own tendency to make systematic errors. They realise they can't depend on their gut to make good decisions. What *feels* right and good won't necessarily turn out right and good. So they don't hide their mistakes and blame others, they examine them and look for lessons to apply the next time.

FURTHER READING

Heath, Chip & Dan, *Decisive: How to Make Better Decisions*, Random House.

12 CONTROL

On October 30, 1935, Major Ployer Peter Hill was testing a new plane when it crashed, killing him and another crewmember. An investigation found that the crash was due to pilot error. Hill had forgotten to release a locking mechanism on the elevator and rudder controls. The new plane had more controls than any previous model and the incident convinced many people that it was too complicated to fly.

Boeing, the manufacturers, couldn't afford to let the plane fail, but they didn't redesign it or reduce the number of controls. Instead, they came up with a solution that probably looked very simplistic. They insisted that all the pilots of the new plane must use a *checklist*. The pilots weren't happy. In fact, most were insulted at being asked to use anything so basic. After all, they were highly trained professionals. They eventually changed their mind when they found that it not only reduced accidents but also speeded up their pre-flight process.

So, the checklist worked, the plane was brought into service and became one of the most famous war machines of WWII. It was the B17 – the Flying Fortress. The USAAC went on to buy thirteen thousand of them and they flew over a million miles during the war without a single accident due to pilot error.

However, the story doesn't end there. The USAAC also began to use checklists for other tasks – for landing and cruising, for bombing and for priming guns. If a task had several steps, it probably had a checklist. The reliance on checklists in the aviation industry continues to this day, where every aircraft has one.

Infections

In 2001 Peter Pronovost, a critical care specialist at Johns Hopkins Hospital, decided to evaluate checklists for doctors. He concentrated on central line infections and persuaded doctors, nurses and administration staff at the hospital to run a trial. After a year they reviewed the results and calculated that the checklists had prevented forty-three infections, saved eight lives and eliminated two million dollars in costs! They were an overwhelming success.

Peter travelled all over the US showing his results to doctors, nurses and insurers. However, few people were interested. Most doctors were insulted that they were being asked to use anything so basic. After all, they were highly trained professionals and any rookie would know what was required. Those checklists were just more bureaucracy. Eventually, in 2003, the Michigan Health and Hospital Association agreed to test his checklist in all state's ICUs. The findings of that test were published in 2006. In the first eighteen months, the hospitals saved an estimated $175 million in costs. More importantly – *there was a saving of more than fifteen hundred lives.*

Surgery

Fast forward to 2006. Atul Gawande, an American surgeon and public health researcher, joined a World Health Organization (WHO) team that was attempting to improve the safety of surgical procedures. People were dying from complications all over the world. The team decided to use checklists to prevent the problems they were seeing. After repeated testing and refining, they eventually agreed on a checklist with 19 items. Most of the items could be regarded as common sense. For example, the first item required everyone in the operating theatre to introduce themselves and state their speciality.

They decided to test the checklist in eight hospitals around the world but were soon facing serious opposition. Many surgeons were insulted and insisted: "Checklists are a waste of time." However, the team persisted, the study went ahead, and the results were finally published in January 2009. Once again, the benefits proved to be enormous. The rate of major complications for surgical patients across all the hospitals fell by 36 percent and deaths fell 47 percent. In other words, *a simple checklist had reduced deaths by almost half.*

Reaction

There was now indisputable proof that checklists saved lives in a surgical setting, as they already did in the aviation and many other industries. Checklists work. Despite the evidence, however, many surgeons were still negative. They complained that the study hadn't clearly established *how* the checklist was producing those results! How is that possible? How can any rational person hesitate to use a simple solution that has such obvious benefits? It seems criminal. Lives are now put in danger because surgeons don't understand *exactly* how a solution works? Can we assume they understand exactly how a defibrillator or heart monitor works?

We often assume that well-educated people have overcome their primordial heritage and harness the logical areas of the brain. These reactions prove it's not true. Education is no barrier to illogical thought (or sheer stupidity). System 1 triumphs over everything. Did System 2 get involved? Yes, but its sole contribution was to justify an illogical stance with a logical-sounding objection. Here's what happened: System 1 made a snap decision – a "gut reaction" – "They're insulting my intelligence." System 2 was then brought in to justify the decision: "We need to understand the underlying mechanism for these results. There could be several reasons…" If they used the same amount of energy to analyse the data objectively before reacting, the outcome would have been so much better.

Checklists and How to Use Them

How do you create good checklists? Atul asked Dan Boorman, an expert on checklist development employed by the Boeing company. He explained that good checklists are precise and easy to use in the most difficult situations. They don't attempt to tackle everything; only the most critical and important steps – those that even highly trained professionals might miss.

A checklist should be as short as possible. One recommendation is that it should contain fewer than 10 items. It should also fit on one page, be free of clutter and use familiar language. It should take very little time to complete, certainly less than 90 seconds, otherwise people will take shortcuts or leave items out. The most critical point is this: no matter how much research has gone into developing a checklist, no matter how many experts have contributed, *it must be tested in the real world*. It must then be revised and tested again and then again.

Types

Before you finalise a checklist, you must define exactly when and where it will be used. What's the trigger? It could be a specific time (9.00am, every Monday), it could be an external event (when our stock price falls below €13.40) or it could relate to a specific task (when the data for the month-end report is available). The trigger helps turn the list of actions into a habit.

But how do you persuade people to use them? The key message that's drummed into pilots from their first days in flight school is that *their memory and judgment are unreliable*, and lives depend on recognizing that fact. You need to provide the same type of focused message for your staff if you want to achieve the benefits of checklists. You need to train people, explain the reasons and sell the benefits consistently, and then maintain that level of effort, because buy-in is very important.

After the WHO results were published, surgical checklists became mandatory in some countries. In Ontario, the reported usage was over 90%, but a follow-up study found "no significant reduction in operative mortality after checklist implementation." The researchers suggested that the reason for the disappointing results was because they were not applied in a "uniform manner".

In other words, just because checklists exist and are *reportedly* in use doesn't mean they *are*. People take shortcuts all the time – that's the whole reason for using checklists. It's not surprising that the same people also take shortcuts when using checklists. That's why they can't just be parachuted in; the culture must change to make sure they're used properly.

Complex and Unpredictable

You can see how checklists can be used to improve the quality and repeatability of repetitive tasks, where every step is known and understood. But what about tasks that are unpredictable? Atul found that checklists could still be used in those situations, but in a different way. For example, in the building industry, he found *"information checklists"*.

Consider the situation: a problem arises and production stops. You and the management team are not available, and nobody is able to make decisions until they contact you. The whole place shuts down. Obviously, that's not right, but what's the alternative? There's no checklist that can deal with *every* eventuality. The answer is to push the decision-making power down the organization. Allow people to make decisions and take responsibility for them.

The *information checklist* determines, for a specific class of problem, who needs to be involved (the team) and how and where to record their decision. It doesn't specify exactly what the problem is, or the potential solution. It doesn't even specify the factors that need to be considered, because there are simply too many variables.

This is an example of *delegation* in action. You're asking a group of professionals to get together and make an informed (System 2) decision. There's no single person making a "gut-decision." And they're also recording the details so you can continue to improve the process itself.

The Wider View

Checklists have enormous potential when properly introduced, but they're not the total solution. In most cases, a checklist is only useful when you have *trained* people available to carry out the tasks. You wouldn't ask an untrained civilian to fly a plane or perform a surgical procedure, no matter how good the checklist. So, a checklist is just the tip of the iceberg. You must also have a *procedure* for the job.

Procedure

A procedure is simply a *method* for doing a job. It's *the way* the job is done. If you have five people each doing the same job a different way, then you have five procedures. However, since there's only one *best* way to do every job, if you have five procedures for the same job, at least four of them are sub-optimal.

Controversial

It's worth pausing here to consider that last point. There's one best way to do every job? Is that true? The idea dates to Frederick Winslow Taylor (1856–1915) and scientific management. While his theories have come in for criticism over the years, there's no doubt that his methods worked. In many cases, it was the implementation of those ideas that caused the controversy. Scrooge-like CEOs often ignored human factors and simply worked people as hard as they possibly could.

The logic is simple. Imagine a job with eight steps and each step takes five minutes. If you can reduce the number of steps to six, then you're saving 10 minutes. Since that time is available for productive purposes, you can increase capacity without increasing workload – everyone wins. It *does* seem very simplistic but in many cases, that level of improvement *is* achievable. The reason is that jobs evolve over time. Checks are added, exceptions are factored in, inefficiencies build up. Like most things, jobs suffer from *entropy* and need to be revitalized periodically.

Writing

Procedures often only exist in the minds of the people who carry them out. When that happens, they're very difficult to improve because there's no guarantee that what's happening today is what happened yesterday. Several people may be doing the job differently and those differences could affect both quality and efficiency. The solution is to document the steps involved – to create a *written* procedure. This can be used as a training document and as a reference for people doing the job. If there's a question about how the job should be done, the procedure acts as the arbitrator.

Many industries – for example, pharmaceuticals and health-care – have no choice about using written procedures, as it's a regulatory requirement. In other cases, customers force suppliers to obtain certification to a specific standard – for example, ISO9000 – which mandates their use. The reason for this level of coercion is that written procedures *work*. They reduce defects, improve quality and improve productivity when applied correctly.

Existing Procedures?

If you already have written procedures (we'll just call them "procedures" from now on) you're probably feeling a little smug. Don't be. Here's why:

- Most of your procedures are unusable
- Most employees can't or won't read procedures
- Most of your procedures aren't used
- Most of your procedures don't describe reality

Unusable Documents

Many procedures are written like legal documents. They use long sentences with multiple clauses. They use jargon, acronyms and complex terminology. They go into long tedious explanations, interspaced with instructions and redundant descriptions. Here's a typical example (try to finish it, and remember, the real thing could run to several pages!):

> *"The vice will need squaring, and this should be done before each job or each sub-task. The work piece on the milling machine is most often held in a vice that is clamped onto the bed and adjustments are carried out to ensure the parts are square and parallel, although it is possible to employ other clamping methods. In that regard, the vice must be aligned with the X travel of the machine. Initially, the vice must be mounted on the bed and secure with T-bolts but this must be done only lightly so as to permit adjustment of the orientation of the vice. A dial indicator is then mounted in the spindle of the machine with the probe facing away from the operator, towards the back of the machine, unless the workpiece is orientated at an angle. The spindle can now be lowered and the bed of the table run back until the fixed jaw of the vice is in contact with the indicator and further until the indicator registers one half of a revolution after which the bezel should them be set to zero and the manual cross feed can be used to run the indicator across the face of the vice..."*

I'll bet you didn't read every word and neither will any worker. If you have procedures like this, you need to get them rewritten and redesigned to look attractive, using colour, photographs and diagrams. Think "magazine" rather than legal document.

People Who Won't Read

Illiteracy is a widespread problem but it's far more likely that many of your staff can read perfectly well and just don't *like* doing it. *Aliteracy* is defined as "the state of being able to read but being uninterested in doing so" and it's becoming more prevalent. In 2016, a Pew Internet study found that 26% of Americans over the age of 16 hadn't read a book in the last year. Some researchers estimate that up to 65% of the adult population suffer from aliteracy. You can imagine what that means for text-heavy procedures.

Procedures should be as short as possible to increase the chances of being read. It can be useful to use three separate documents: a *checklist* for essential tasks, a *procedure* for instructions, and a *training manual* to describe how the process works, how things are laid out, equipment details and other non-critical information.

Procedure Problems

Some companies have shelves stocked with procedures for every task and every piece of equipment. In theory, they're available to anyone who needs them, but of course, few do. It'd take time and effort to find the right procedure and figure out what it says, so people just figure it out for themselves. Those procedures are only useful when customers or auditors come around.

Often a procedure only describes what's happening on the day it's written. Changes occur but the written procedure isn't updated. That means the entire procedure is useless. Why? Because there's no way of telling which parts are wrong.

Procedures need an *infrastructure*. There needs to be a formal way to change them and to make sure the current version is always available. More importantly, there must be an enforcement mechanism. There are two ways to achieve this:

- Become certified to an external standard e.g. ISO9000
- Enforce your own internal system

External Standards

When you're externally certified, it becomes easier to justify compliance ("we'll lose our certification unless everyone follows the procedures"). It'll also be easier to justify dedicating resources to run and enforce the system. The external agency will carry out periodic audits to make sure you're compliant and they'll pull your certification if you don't follow the system. The problem with an external standard is that it may force you to increase the level of administration significantly. You might have to generate procedures for non-critical items, and inspectors may insist that you monitor some areas more closely and keep records of those checks.

Internal System

When you enforce your own system, you're in total control and can decide what to include and exclude. However, enforcement can be problematic. The only way the system is going to work is to consistently prove to everyone that the system takes precedence over everything else. For example, suppose your entire production process has stopped because a machine has broken down. You can run the product on another machine but you've no procedure for it. It'll take at least three hours to write a new procedure. What do you do?

This is the type of dilemma facing team leaders every day. Their choice will be informed by how *you* react. If you insist that they should have changed over immediately without a proper procedure, you've made an expedient choice, *but the use of procedures is dead.* Anything you say afterwards will fall on deaf ears. Procedures impose a discipline on an organization. They reduce chaos. They ensure that things are done in a predictable and repeatable manner. But they only work if everyone takes them seriously, particularly management – specifically, you.

Management Conflict

Every manager wants to keep the operation running, particularly when disruptions occur, and decisions must be made. Any default decision is likely to be based on short-term consequences: "I won't make my figures unless I get the other machines working. I'll worry about the procedure later." They may not see any real downside – that's your job. You must explain that anything they'll gain by taking shortcuts is a drop in the ocean compared to the long-term losses suffered because procedures aren't being followed. You must take the hard decisions.

Quantity?

How many procedures should you have? The *least* number that'll ensure your key processes are protected. The 4-1-1-4 tool illustrates what you should be aiming for:

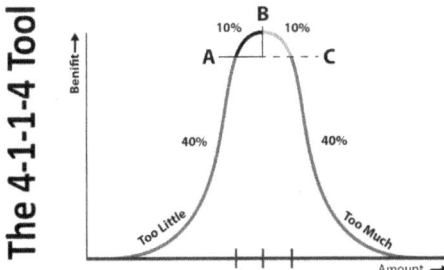

The ideal number of procedures is between A and B on the curve. Any fewer than A and some of your critical processes aren't being controlled properly. Between B and C, you're getting the benefits but you're putting in more effort than necessary. More than C and you're suffering from an administration burden. Remember, procedures must be kept up-to-date and changed when necessary. They must also be checked to make sure they're being followed. The more you have, the greater the administration overhead.

Checklists and procedures are the best way to ensure that people are doing their job in the right way. They also provide the only way of making systematic improvements to the process. We've seen that you're *personally* responsible if anything goes wrong. Procedures are your control rods to make sure that doesn't happen.

Improvements and Innovation

If you depend entirely on checklists and procedures, you'll eventually end up with many instances of *functional stupidity*. People will do their job, and do it well, but they won't see the big picture. The result will be waste and duplication. That's why you need four additional components:

- A review system for people using the procedures
- Change control guidelines
- Feedback from downstream and upstream areas
- An outline of the organization's objectives

A Review System

The people using the procedures must have the opportunity to *change* them. They're closest to the process and most likely to see opportunities for improvement. If a change is proposed from an outside source, they must be involved because they may see problems an outsider could miss.

Change Control Guidelines

It's not enough to allow people to change the procedures; they must have enough *knowledge* to do it in a safe way. That means they must have change control guidelines. For example, it might be possible to speed up a process by increasing the operating temperature, but this could cause reliability issues. The guidelines will point out and prevent issues like that.

Feedback

When people focus on their own part of the process, they may improve it but cause problems in other areas, or for the final customer. That's an example of functional stupidity. The best way to prevent it happening is to give them *feedback from upstream and downstream areas*. They need to see where they fit into the total picture. That knowledge improves the chance that changes will benefit the entire organization. They should also be willing to take a small hit in productivity (for example) if it results in large gains in another area. If that happens, it's important they're not penalized – the metrics must be flexible enough to reward such innovation.

Outline the Organization's Objectives

Sometimes people work to improve a process only to find – too late – that a management decision has been made to shut it down and move to a different process. That's a waste of resources and always leads to frustration. People are less likely to try to improve the next process. The solution is to make people aware of any potential changes that could affect their area.

FURTHER READING

Gawande, Atul, *The Checklist Manifesto: How To Get Things Right*, Profile Books.

13 STORIES

On Tuesday 15th March 2005, Ann Coulter received her first order. Her new company, NETRIPUM, located in Sacramento, was building an advanced network integrator for medium-size IT departments. The order was confirmation that she was on the right track. Another company called TASRETS had the same idea. This were located just outside San Diego, and run by former NASA engineer Daniel Farrier. Finally, in Richmond, Virginia, a third company had opened its doors. The company, NET-ZOUP, was run by Brandon Toste, a telecommunications engineer, and they were also competing in this market.

It may seem more than a coincidence that three companies would set up within months of each other and produce essentially the same product, but it's more common than you might imagine. It happens when the underlying technology has matured to a point where the next innovation becomes almost inevitable. People are just waiting in the wings to seize the opportunity.

Each started with approximately $1M, which they received from banks and venture capitalists. This amount was a function of the perceived market potential and risks involved. However, even though the three entrepreneurs started in the same way, the outcomes were very different, as can be seen from the following graph:

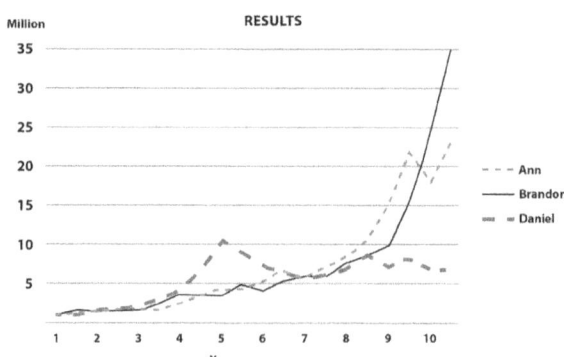

Brandon's company was worth \$35M at the end of 2016, Ann's was worth \$24M and Daniel's less than \$7M. What could have caused similar companies to end up with such different results? One journalist – Adam Greentree – began to investigate. First, he examined the *number of products* that each company supported. These included different variations of the same product. For example, a customer might require a higher bitrate on a specific data channel. A version with this change could then be offered as a stand-alone product.

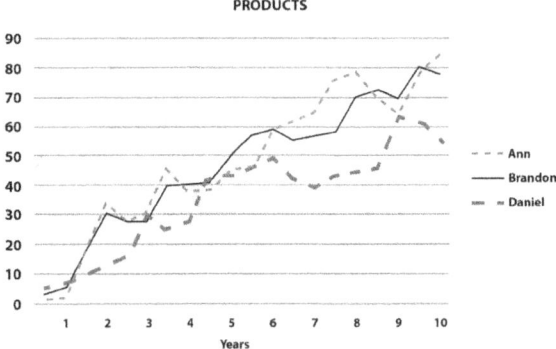

Adam found that Daniel (the least successful of the three) almost always offered fewer products than Brandon and Ann. This was particularly noticeable after year 6 (2011) when the company had suffered a downturn and was struggling with poor sales. Ann's results tracked Brandon's reasonably closely until year 9, when they suddenly dropped. The following year there was also a significant drop in her sales. Next, Adam looked at the number of salespeople employed by each company.

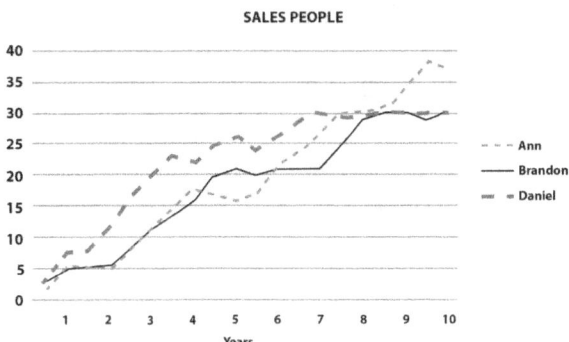

He found that Daniel (the least successful) had employed significantly more salespeople than either Ann or Brandon until year 8 (2013) This could suggest a dependency on face-to-face sales as opposed to advertising and web-based sales. Ann and Brandon's figures tracked reasonably well until year 8, after which Brandon kept his sales force constant while Ann's continued to grow. Brandon had realised that alternative sales methods were now becoming more effective. Ann, however, didn't learn that lesson for another year. Finally, Adam found a clear trend when he looked at the total staff numbers.

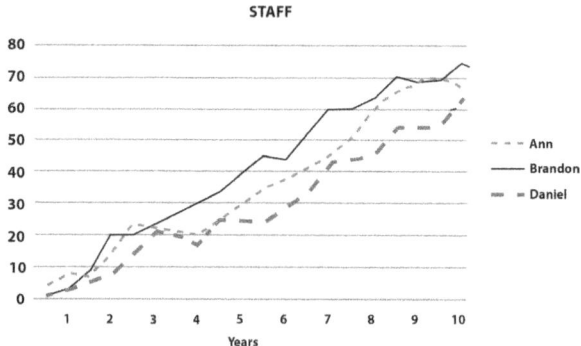

Brandon (the most successful) always employed more people than either of the other two (except for a few months in year 2). This is particularly relevant because he also had fewer salespeople. Network integration is a highly technical business and Brandon's additional staff give him the flexibility to develop new products while still supporting non-routine customers' requirements. Adam summed up the situation as follow:

"Brandon Toste has built NET-ZOUP to become the acknowledged leader in the medium-size network integration business. His foresight and steady handling have meant that the company has grown consistently and avoided the pitfalls that have befallen his competitors.

Daniel Farrier led the pack for five years, but he made the mistake of cutting back on staff at a crucial time. He compounded this by building a large salesforce when other sales channels were becoming more effective. Once he lost the lead in 2011, he was never able to recover the initiative.

*Ann Coulter's company has been a consistent performer but has stum-
bled in the last few years. This is due to several factors, including a
sharp reduction in her product range in 2014. She continued to increase
her salesforce when a more conservative position would have been
prudent. The additional costs have limited her options in other areas."*

Your Review

Based on the information above and your own experience, please answer
the following questions:

On a scale from 0 (unsure) to 9 (absolutely certain), please rate the
following statements:

- Ann made a mistake cutting the number of products in year 9
- Daniel shouldn't have expanded his salesforce as fast as he did
- Ann would have done better if she hadn't cut back on her
  workforce in years 3 and 4

On a scale from 1 (poor) to 9 (excellent), rate the following:

- Brandon's decision-making ability
- Daniel's decision-making ability
- Ann's decision-making ability

On a scale from 0% to 100%, rate the following:

- How much of Brandon's success was due to chance?
- How much of Daniel's loss of leadership was due to chance?
- How much of the difference between Brandon's
  and Ann's results was due to chance?

Finally, if you were to select one of those CEOs to lead a new company
with the intention of selling it on *within five years,* which CEO would
you choose?

> **Please don't read on until you've written down your answers to all
> those questions (Q13-1).**

The Perception of Chance

This section looks at your perception of chance and what it means for decision-making. If you read newspapers or business magazines, you'll have seen articles describing CEOs and their organizations in similar terms to Adam's descriptions above. If you read business biographies or autobiographies, you'll come across similar histories and analysis. How do you react to that information? How much of your business knowledge is built on those stories and does it make it easier or harder to make logical decisions? These are the questions we'll be investigating. But first I have a confession. *Adam doesn't exist and neither do Ann, Daniel or Brandon, or their respective companies.* I apologise for the deception but, as we'll see shortly, there's a very good reason for it.

Data Review

Let's start by looking at the data. All of it was created randomly by rolling a dice. The first set of numbers – the "results" were created by rolling the dice 20 times for each "CEO" and then applying a simple formula to the results. The same formula was used in every case, so *any differences in the results are due only to chance.* The same is true of the products, salespeople and staff data. That means:

- There were *no decisions* made to achieve those results.
- There's *no connection* between the results and supporting data.

In other words, Brandon didn't make any good decisions and Daniel and Ann didn't make any bad ones. All the results were random. So what's the point? The point is whether you accepted the results and the story around it and whether you were influenced by "Adam's" interpretation of the data.

Look at your answers and see what they're telling us. For example, how certain were you that Ann had made a mistake cutting back on the number of products? You *should* have scored zero, since there was *no* relationship. The higher your score, the more you were influenced by the jottings of a fictitious journalist and a coincidence between two random data sets. How did you rate each CEO's decision-making capability? Again, it should have been zero in each case (since they didn't make any decisions). The higher your score, the more you were influenced by random variations in the data.

The next set of questions is interesting because you were asked to consider the role of *chance* in each scenario. You should have scored 100% for each question but it's likely you didn't. The lower your scores, the more you were seduced by the *illusion of causality*.

Finally, you were asked to select a CEO. You weren't told how to do it, but it was inferred that you should use the data available in the graphs. Of course, the data had no bearing on the result. If you made a selection, it shows that your decision-making capability is satisfied by readily available information. The right answer? "Insufficient data."

Unfair!

You probably think that exercise was unfair. After all, you were asked to review the information in good faith and now you find that the facts and data were fabricated.

In fact, that isn't quite true. The data *weren't* fabricated, they were generated randomly, and that distinction is an important one. Before the first die was thrown, nobody knew which "CEO" was going to win and by how much. Nobody knew what differences would show up in the metrics. All the key information was directly derived from that data, so let's recreate the mind of a pundit looking through it, trying to find a reason why one company is successful while another isn't.

Obviously, *the first thing they'll identify is who's been successful.* In this case, Brandon is the clear winner with Ann coming in a good second and Daniel bringing up the rear. So what's the reason for the difference? There were four sets of data available:

- number of products.
- total staff.
- salespeople.
- partners.

The "partners" data show no trend, but the staff data show a clear difference between the leader and everyone else. That *must* mean that

Brandon has the right number of people, and the others are understaffed. So, there's our first "reason" identified.

But suppose things had worked out differently? Suppose Brandon had the smallest number of staff? In that case, he managed his costs better than the others and *they* wasted resources unnecessarily. If there's a pattern, we can find a story to justify it – that's what we've done here.

Let's take another example. Daniel was ahead until year 6 when his results dropped away. What happened? Nothing much shows up on two of the charts, but the staff numbers dropped significantly in year 4 so that's *obviously* the reason. And so, we build a story around that. It easy to assume that one thing is related to another when both changed around the same time. But, in this case, we know that's a false assumption because there's no relationship. Not just no causality, *there's no relationship whatsoever* – they're not connected in any way.

You might still be arguing that the exercise was unfair – there was no way you could have known that the data had no relationship to the results. *But that's the point.* You're making decisions every day using "data" that have little or no relationship to the outcomes you're observing.

Biases in Action

There are many physiological processes at work here. First, there's *survivor bias.* We assume there's something special about an individual or company because we can see their success *today.* However, that's not necessarily a good indicator. For example, if we carried out a review in year 5, we'd think that Daniel was the best CEO by a significant margin. If we carried it out in year 9 we'd be certain that Ann was the best. Who knows what might happen in year 12? And, of course, we know that none of them are really special!

The second process to consider is the "*what you see is all there is*" bias. In other words, you were given a certain amount of information and you used it to make decisions immediately. You didn't stop and say, "Wait, I need more information before I can decide. This isn't enough!" You must *always* ask yourself what's missing.

The third process is the "*authority bias*". It's likely that you accepted what Adam, the fictitious journalist, was telling you without checking everything for yourself. For example, did you notice any discrepancy in his explanation? What about that drop in staff numbers in year 4 leading to a drop in earnings in year 6? Even though year 5 was the best year by far? That's not impossible but it's highly unlikely. And yet you probably let it slide by.

If you listen carefully (and critically) you'll often hear people repeating exactly what some commentator said on TV the night before, as if it was their own opinion. When you hear the same thing from multiple sources, eventually you'll assume it must be true. The constant repetition can easily break down any reservations you have about accepting it – that's how propaganda works. And that brings us to *stories*.

Stories

We tell stories all the time. We even tell ourselves stories; it's how we make sense of the world. Have a look at Daniel's results again:

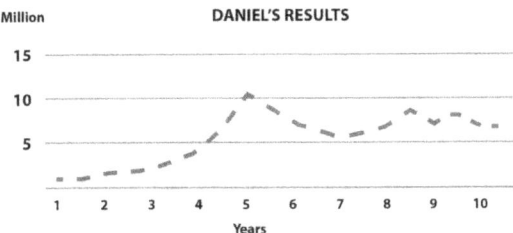

You can see that something happened in year 6 and he never recovered from it. If you were told he went through a messy divorce, started drinking and he's been in and out of rehab since then, would you tend to believe it? Of course, Daniel doesn't exist, and nothing happened in year 6 but the story is compelling. It explains a lot.

Emotional Appeal

Many of the decisions you make are probably based on similar stories. They have an emotional appeal but no data to back them up. Just think about where you got your beliefs about management. You have your own experience, but you've also listened to people talking about their experiences, either as managers or as employees. You've also read stories about great CEOs, perhaps even some autobiographies? *Unfortunately, they're all lies.* OK, perhaps it would be more correct to say they're all untrue (but not as dramatic or emotional!). Let's consider a short story (from Jack J. O'Neill's autobiography):

> *"It was in February of 1985 that I decided to get into the electronics business. I set up a small shop in Bromley, London and started looking for products that would sell. I went to the States and brought back an entire shipment of Pocket-Vision TV sets from Radio Shack – there must have been at least 100 sets. I placed advertisements in the Sunday national papers and they started selling immediately. They were all gone within a week. That's when I knew that I had a winner, but I needed something bigger, something with a theme. I was the first shop in England to stock the Sinclair ZX Spectrum. When that started flying off the shelves I got rid of everything else and bought huge shipments of the Amstrad CPC464 and the Acorn BBC Model B..."*

It sounds like this entrepreneur had a clear vision. He was always destined for greatness. However, the reality was somewhat different. Let's look at some of the reasons why that story isn't *exactly* true:

Lies

The shop was owned by Alan Greyson, his uncle. When Alan died, the shop was left to him. He never went to the States. He asked a friend to ship the TVs back.

Omissions

He'd been working for Alan for three years before he was left the shop. The money for the stock was provided by his mother. She never got it back.

Remembered Incorrectly

He placed ads in the local weekly papers, not the Sunday nationals. It took over three months for the TVs to be sold. He wasn't the first shop to stock the Sinclair – it had already been around for three years. One of his friends, George Feder, persuaded him to get into computers. He didn't get rid of the non-computer items until 1989.

Exaggerations

The first shipment contained 15 TV sets – not 100. He never had more than a dozen computers in the shop at any time

Distortions

The shop was on the edge of collapse throughout 1985 and 1986. It was only his mother's money that kept it afloat during this period. It was saved because a local firm pre-ordered 50 IBM 5150s. The order was placed only because the firm was located directly across the road.

Deterministic Systems

That illustrates the difference between story and reality. Do you feel the same way about Jack after reading the second account? Whenever you hear a story, you use it to update your model of the world. However, if the story isn't true, your model of the world diverges from reality.

This example also illustrates an important point. *The world is not a deterministic system.* A deterministic world would always produce the same output from a given starting point. *There would be no randomness involved.* However, we know that's not true. For example, if the company across the road wasn't looking for computers at exactly the right time, the results would have been very different for Jack.

Probabilistic Systems

If the world is not deterministic, what is it? It's *a probabilistic system.* You can't say for sure that something *will* happen, but you may be able to estimate the *probability* that it'll occur. For example, you can't say for sure that it'll rain tomorrow but you can calculate that there's an 80% chance it will.

What that means is: when taking all the information we have available (like temperature, pressure, satellite imagery, etc.), it rained 8 out of every 10 times a similar pattern occurred before. We used this approach in the forecasting chapter but what does it mean for decision-making? Let's look at three statements you might make if you believe in a deterministic world, and translate them to reflect reality:

We're going to upgrade all our computers to the new OS.

This statement describes a situation where you have a reasonable level of control. You have the computers, the software, and the people to do the job and you don't have a time limit. It looks *certain* that you'll be successful. However, you did specify *all* the computers. It's possible that some users will be particularly resistant, or some essential software packages will be unable to run on the new OS. In that case, it might be impossible to achieve the goal. Something that initially looked certain (100% probability) could easily become impossible (0% probability).

One way around this might be to specify a certain percentage. For example, 95% of computers will be changed over. However, even this might run into difficulties. Some people might be off sick, other people leave, priorities change, and the project is put on the back burner, never to be completed. If you've created a reactive culture, the probability of this happening is more likely. History becomes the best guide in these cases. An impartial analysis might lead to an alternative prediction: *There's a 75% probability that we'll upgrade at least 95% of our computers to the new operating system.*

The new machine will increase throughput by 25%.

This statement is more difficult to quantify. You don't have the machine yet and you've only seen limited tests. Those looked good, but things could be different in production. Let's look at some probabilities:

- The machine will pass acceptance testing: 94%
- It can be maintained to run at maximum speed: 92%
- The workers will be trained to maintain run rates: 91%
- The machines will run at least 16 hours every day: 90%
- Downtime will be less than 2 hours a week: 90%

Those individual probabilities look reasonable, but they must be multiplied together to get the overall probability of achieving the result. So, here's the corrected statement: *There's a 64% probability that the new machine will increase throughput by 25%.*

The marketing campaign is going to increase sales by 2%.

This is more difficult because you have no direct control over the most important part of the project – the reaction of the customers. You can try to minimize the risk by carrying out pilot studies, but there's still a lot of uncertainty involved. Let's break the project into two parts: Part 1 where you have a good deal of control and Part 2 where you haven't.

Part 1 includes all the activities that are under your direct control, including developing a plan, designing and filming advertisements, printing fliers and buying media space. There are a lot of individual tasks, so while individual probabilities are in the high 90s, the overall probability that everything will be completed as expected is 74%.

The second part is more difficult to estimate. One solution is to take the external view and look at base rates, where you'll find the following: Of 100 similar campaigns conducted over the last five years:

- 29% met all expectations
- 22% did not meet all goals but were deemed "acceptable"
- 31% were deemed unsuccessful
- 18% were regarded as a disaster

If we're willing to accept "acceptable" as our measure of success, then we have a 51% probability of being successful in Part 2. This must be multiplied by the probability of success in Part 1 to get the overall probability. That means our prediction should read: *This marketing campaign has a 38% chance of increasing sales by 2%.* Perhaps not as positive as we hoped?

Does it Matter?

Let's say you must decide between the following two projects:

- This campaign is going to increase sales by 2%
- This campaign has a 38% chance of increasing sales by 2%

You'd be more likely to pick the first one, wouldn't you? Yet we know that both statements describe the same project. The second description gives you more information, allowing you to make a better decision.

Now, select between these projects:

- This marketing campaign will increase sales by 2%
- This marketing campaign will increase sales by 3%

Which one would you pick? Obviously, you'd go for the second. However, if you discovered that the probability of success in that case was only 12%, you might change your mind. The additional information makes all the difference.

Stories and Parables

The stories in this chapter are parables – they're not meant to be mistaken for the truth. But, by now, you should realise that *all stories are parables*. They should never be regarded as anything else, because stories can *never* be totally true. That means our view of the world is based on partial truths, untruths or lies. Even the memories of our own experiences are an edited version of reality – a distortion that our System 1 is comfortable with because it rarely questions a good story. The only solution is to force ourselves to use System 2 and treat the world as it is – a probabilistic system. There are no certainties. After all, how can you be sure what's going to happen in the future when you can't depend on your own memories of the past?

Decision-Making

Our tendency to concoct stories to explain random events is called "*confabulation*". It's defined as the production of fabricated or distorted memories about the world without a conscious intention to deceive. For example, an employee notices that you've come in late for work three mornings in a row. That's unusual, so she immediately concludes that you're looking for a new job. Before long, the story has spread throughout the office. Gut decisions are based on stories – your own and those you've heard or read about. That means they are based on a distorted view of the world. So, one last story:

Margaret Longton wanted to invest in a small store in downtown New York, but the rents were shocking, and the previous tenants had failed miserably. She hired a consultant to review the options and he came back with the prognosis that her chance of success was around 10%. She went through his logic and couldn't find a flaw. Nevertheless, she had a strange feeling that things would work out. She went ahead and opened the store and it was a success. Was her decision correct?

> **Write down your answer before reading on (Q13-2).**

Before simply accepting this story – and similar ones – as a learning tool you should add the following:

1. She opened the store and it was a total failure.
   She went bankrupt.
2. She opened the store and it was a total failure.
   She went back to work in a retail shop.
3. She opened the store and it was a total failure.
   She suffered from a nervous breakdown.
4. She opened the store and it was a total failure.
   She was hospitalized for malnutrition.
5. She opened the store and it was a total failure.
   She was in serious debt for the next twenty years.
6. She opened the store and it was a total failure.
   She decided to become a hippy and live off the grid.
7. She opened the store and it was a total failure.
   She was plagued by bailiffs who seized her belongings.
8. She opened the store and it was a total failure.
   She had to live in the basement of her mother's house.
9. She opened the store and it was a total failure.
   She left the country and became a nun in Mexico.

That's right, you hear about the successes, but you rarely hear about the failures, and *you never hear about all of them.* The "success against all advice" story is a classic that gives people a highly distorted view of reality. Once again, we're fooled by *survivor bias.* The outcome is never a good way to judge whether a decision is correct or not. So, what *should* she have done?

Write down your answer before reading on (Q13-3).

You know the answer by now: *she should have walked away*. The odds were against her. That's the difference between a professional gambler, who calculates the odds and makes money, and the amateur, who believes her gut and (usually) loses it. It's simply a matter of professionalism.

FURTHER READING

Dan Burnett, *The Idiot Brain. A neuroscientist explains what your head is really up to*, Guardian Books.

14 PRICING

Setting the right price for your product or service is one of the most important things you can do. That decision will determine who your customers and competition are going to be – and it'll strongly influence your earnings. Despite its importance, many CEOs don't spend enough time investigating all the options. As a result, they forsake a wonderful opportunity to create higher profits for their organization. Let's begin by considering two common pricing strategies: the "cost-plus" and the "follow-the-leader" models.

Cost-Plus

The cost-plus model works by adding up all the costs of making the product and adding a percentage for profit. It sounds easy, but there are complications, because there are *two* types of cost: fixed and variable. *Fixed costs* are those that would exist even if the product wasn't being made. They include rent, buildings, machinery and similar costs. They usually don't change as the volume increases or decreases. *Variable costs* are those that change in proportion to the number of units produced. These include wages, direct materials, utilities and consumables.

Let's take a simple example. Suppose you're going to launch a new product. You've calculated that the variable costs are €10 per unit and the fixed costs are €100,000. You want to make a profit of €5 per unit. How much should you charge? You can see that the cost of each unit will depend on the number of units sold because your fixed costs need to be apportioned. You can't calculate the price until you know the volume. If you're likely to sell 100,000 units then the cost per unit is €11 and you should charge €16. However, if you're only likely to sell 10,000 units the cost per unit is €20 and you need to charge €25. That's a big difference – and all due to the volume of units sold. The trouble is, you won't know that figure until *after* you launch the product, so you could be in trouble if you miscalculate.

Follow-the-Leader

The "follow-the-leader" model takes a different approach. Instead of calculating the costs and adding a profit, you simply look at the price that the market leader is charging and then set your price at a certain percentage of that. If you want to compete on price, you might set the percentage to 90%. If you want to compete on quality, you might set it to 120%.

There are advantages and disadvantages with both approaches. In the first case, you're factoring in your costs, so you *shouldn't* lose money. However, you're not considering what the customer is willing to pay. You might be charging too little, in which case you're leaving money on the table, or you might be charging too much, in which case you're losing potential customers.

There's also a danger when you're following a market leader, because she may have a different cost structure to yours. She may be buying raw materials at a lower price or she may be more efficient at manufacturing the product. Even if her costs are the same, she could be selling more units – so her cost per unit could be significantly less. For any of those reasons, you could end up selling at a loss.

The Demand Curve

Those models don't take into account how the customer is going to react. However, the price you charge will influence the number of people who're going to buy. For example, how much would *you* pay for a chocolate bar – assuming you like chocolate? It's probably safe to assume you wouldn't spend €100 on one? Probably not €20 either? This shows that certain price levels will *prevent* people from buying the product. Let's say the most *you'd* spend is €5 but other people might be willing to spend up to €9.99. You need to take this variation in people's willingness to pay into account when you set your price. The demand curve is one way of doing this:

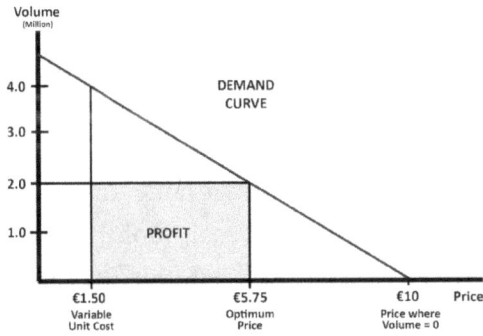

The price on the right (€10) is the price at which the number of units sold is zero. Nobody – not even the most ardent lover of the product – is willing to pay that price. The price on the left (€1.50) is the variable cost of the unit. Below this point, the price doesn't cover the cost of making the unit. The optimal price (€5.75) is halfway between the variable unit cost and the price at which the number of units sold is zero. The volume at the optimal price is 2 million units. The profit is shown by the shaded rectangle.

Price Elasticity

Some items are more sensitive to price changes than others, but how do you measure that characteristic for *your* products or services? You'll need to understand the term: "price elasticity". That's *the percentage change in sales volume divided by the percentage change in price.* Let's take an example. Suppose you *increase* your prices by 20% and the number of units sold *decreases* by 30%. In that case, the price elasticity is 30 divided by 20, which is 1.5 (the volume is going down but we can ignore the negative sign). Some goods, like petrol, cigarettes and electricity are inelastic. The volume changes very little with changes in price. For example, electricity prices might go up by 10% but usage might only drop by 2%. That means the cost elasticity of electricity is 0.2.

Other goods, like Heinz beans, Cadbury chocolate, and yachts can be very elastic. The reason is that it's easy to switch to a substitute – or in the case of the yacht (which is just a luxury) – avoid the expenditure altogether. If the price of the yacht increases by 10% the volume might decrease by 30% – a cost elasticity of 3!

Using Price Elasticity

You can use the price elasticity of your product or service to help set a price that'll result in the highest possible profit. Let's have a look at how this works: Suppose you're selling haircuts for €20. The fixed costs are €10,000 per month and the variable cost is €5. Together with your staff, you give 1000 haircuts each month, so there's a fixed cost contribution of 10,000/1000 = €10 per haircut. That means your profit is €5 per haircut or €5000 per month. That's not bad, but what would happen to your profit if you were to increase the price? What does your gut tell you?

Write down your answer before reading on (Q14-1).

Let's say you increase the price to €24 and the number of haircuts drops to 700 per month. In percentage terms, you've increased the price by 20% and the volume has decreased by 30%, so your price elasticity is 30/20=1.5. What happens to your total profit? It goes from €5000 to €3300 a drop of 34%. Not good, and if you'd increased the price to €30 it would have been worse – you'd have seen a loss of €3750 because the volume would drop to 250 haircuts.

What would happen if you *dropped* the price? If you dropped it by 20%, you'd be giving 300 more haircuts every month, but your overall profit would drop by €700 to €4300. So it looks like €20 is the right price? It's close, but not the optimum. If you were to drop your price to €19, you'd make €5050. That's not a huge increase, but every bit helps!

Competitors

If there were no competitors your job would be much easier but there are, and they're trying to grab *your* customers. If they dropped their prices, how would you respond? If you maintain your prices, you'll lose some of your customers and if your product or service is very elastic, the effect could be significant. As we've seen, *the decrease in volume will effectively increase the cost of every unit you produce,* so your profit will shrink or even disappear. On the other hand, if you drop your prices, you might maintain your *volume,* but you'll still have less profit. You can calculate the impact of a competitor's action using the "cross-price elasticity". This is the percentage change in *your* volume divided by the percentage change in the *competitor's* price.

If you find yourself in a situation where your competitor has dropped prices to gain market share, you need to stop and think very carefully. The obvious (System 1) reaction to also drop prices could trigger a price war that could eat into your profits. You need to calculate the effect of *every* possible response before you act. You face the same dilemma when *you're* considering dropping your prices to increase market share. Your competitors may respond in kind and, if they do, all you've done is reduce the profitability for everyone in the industry – including yourself!

Single v Multiple Prices

A single price is elegant, isn't it? No messing around with multiple price lists and different customer segments. Unfortunately, it also leaves a lot of profit on the table – something you really can't afford to do. Let's look at the demand curve again:

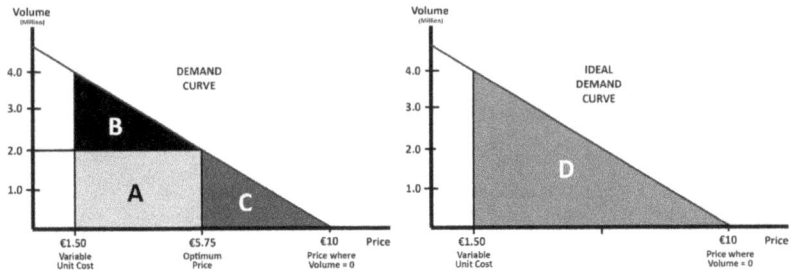

We've seen that the profit rectangle (A) is contained between the variable unit cost and the price. What about the triangle (B) over the rectangle? You're getting no profit there because potential customers aren't willing to pay the price you've set – it's too high. You'd have to drop your price to get this additional volume. What about triangle C? The people in this area *would* be willing to buy your product at a higher price. However, they don't need to because they're already getting it at €5.75.

So, even though you've optimized your price, you've left a lot of profit behind. The ideal way to get that profit is to set an individual price based on what *each* person is willing to spend. That would result in the graph on the right where all possible profit has been realised (D). That's not usually realistic – but suppose you were to set three prices? You could set the first at €3.62, the second at €5.75 and the third at €7.87. The graph below shows the results. It turns out that this approach does allow you to reclaim some of the extra profit (B and C).

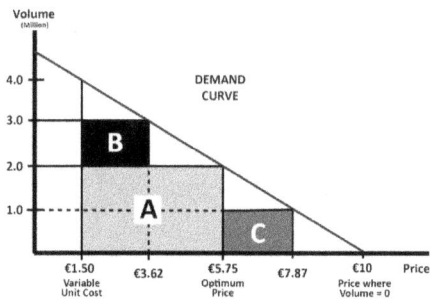

But is this really going to happen? Why should I pay €7.87 for a product when I can get it for €3.62? That's only going to work if you can invent some mechanism to prevent high spenders from selecting low-cost options.

Fencing

So, how do you charge *different* prices for the *same* product or service? The answer is to fence in your prices. You create a situation that prevents the person willing to pay €7.87 from being able to pay €3.62. That can take ingenuity, but it *is* possible. For example, supermarkets used to sell their own brands in bright yellow packaging. These "yellow packs" were designed to be highly noticeable at the checkout. Why? Because it was a visible sign that you couldn't afford branded products. If you *could* afford them, you'd stick to the more expensive products to avoid the stigma of the yellow packaging.

In 2000, Škoda, the Czechoslovakian car maker, became a wholly owned subsidiary of Volkswagen. At the time, Škoda cars were the butt of jokes because of their unreliability and tendency to rust. VW invested heavily in the company and increased the quality to Western standards. Within a few years, they were producing cars that were every bit as good as those produced in Germany. However, the Škoda reputation lingered.

When VW bought the company, they agreed to the Czech government's request that the Škoda brand name be maintained. It was consistent with their strategy. Škoda was the "yellow pack" of the car world. Its reputation allowed VW to sell what was essentially the same car at lower prices without impacting their existing sales. That resulted in a significant increase in profits.

At the other end of the scale, VW own the Audi brand. This is a luxury brand for people who wouldn't consider using an "ordinary" VW. They are willing to pay more for the prestige of owning an Audi even though, again, they're buying essentially the same car. Those examples show how "*snobbery fencing*" can be used to optimize profits. There are other types of fencing that you might also consider:

Time fencing

Airlines charge more for flights at certain times. The most expensive days are Friday and Sunday. The cheapest days to fly are Tuesday, Wednesday and Saturday. The cheapest times to fly are when most people don't want to – for example: dawn, late at night (red-eyes) and flights around lunch and dinner hours. People who can afford it will spend more to avoid those times. Hotels and restaurants use similar tactics.

Geographical fencing

You can buy a CD or DVD in certain countries at a fraction of the price in others. Media companies have established a different demand curve for each territory.

Bundling

Another way to increase profit is to offer a bundle of products or services, rather than just individual units. Suppose you're offering broadband, fixed and mobile phone services.

The stand-alone monthly price, subscribers and total revenue are:

		Subscribers	Total
Broadband	€ 30.00	100000	€ 3,000,000
Fixed	€ 25.00	30000	€ 750,000
Mobile	€ 35.00	120000	€ 4,200,000

Total	€ 7,950,000

Now, if we offer a bundle that includes all three services for a single fee of €50, and just 10% of your subscribers from each service switch over, then you'll get the following:

		Subscribers	Total
Broadband	€ 30.00	90000	€ 2,700,000
Fixed	€ 25.00	27000	€ 675,000
Mobile	€ 35.00	108000	€ 3,780,000
Full Bundle	€ 50.00	25000	€ 1,250,000

Total	€ 8,405,000
Increase	6%

That's a 6% increase in revenue without any changes to the services provided. Of course, there's a danger that people who previously subscribed to the three services separately will subscribe to the bundle, so you need to research your customer subscriptions and the options you're about to offer very carefully.

Anchors

So far we've looked at the "logical" side of pricing. Now we're going to look at something else. Let's say a company is offering a completely new product. They start by offering it at €299.00. You see it advertised at that figure for several months. Then you see a new version with most of the same features, and it's being sold for €99.00. Would you buy it?

If you were interested in the product, you'd probably find the offer irresistible. The reason is a process called *anchoring*. The initial price sets your expectation as to how much the product is worth and the lower price appears to be a bargain. But suppose they *initially* offered it at €99.00 and the next version, with essentially the same features, was sold for €299.00? The anchor would now tend to drag down demand. You're being asked to buy something that you think is worth less than they're asking for it.

An anchor can be very powerful and can have unintended consequences. For example, if you launch a new product and want to capture market share, you might offer it at a special low introductory price. However, that price could become an *anchor* – so any attempt to return to the "normal" price will be resisted.

The Rule of Three

Do you think it's a good idea to sell a product that nobody wants or buys? It would seem to be sheer lunacy, but there's a reason you might want to try it. Let's say you sell two products. The basic model (on which you make very little) sells for €300, and a more advanced model sells for €500. You want people to buy the advanced model, but only 20% of your sales are currently coming from it. Instead of dropping the price (the natural reaction) you could try introducing a new model for €750. Even if nobody buys it, it's likely to increase sales of your €500 model *because people tend to choose the middle product when given a choice of three.*

Scarcity

Sometimes you can increase sales by *limiting* how many units you produce. If people believe a product is scarce, they're more likely to buy it quickly and may even order several. For example, if you're shopping on Amazon and see the notice "Only one left in stock", are you more likely to order it? If you're shopping for hotel rooms and see "Only one room left at this price", are you more tempted to buy right away? That's scarcity in action. People will pay less attention to price if they're afraid they might lose out.

Companies often use the threat of scarcity to raise prices but sometimes that can be counter-productive, because people regard it as *price gouging*. In that case, the reputation of the company is sullied, and it can have a negative effect on future sales.

Displaying Prices

There's been a lot of research on how the brain reacts to *money*. One interesting finding is that price information activates the brain's pain centres. In other words, just looking at prices hurts! The pain is worse when viewing prices displayed using common formats – for example: €22.00. Currency symbols seem to be particularly painful. The *least* pain is experienced when the number is written in letters without a denomination: "twenty-two" in the example above.

Pricing Considerations

You'd expect to pay more for a luxury brand than an "ordinary" one, but this expectation works both ways. People often believe something is a luxury item *primarily* because it's more expensive. The high price implies that the item is better than its "ordinary" competitors. A price discounting strategy could be devastating for those brands. Imagine a high-end perfume that normally retails at €1500 an ounce suddenly being discounted to €99.99 a bottle!

Pricing Mechanisms

We've looked at some basic pricing mechanisms here, but there's much more to explore. The five key things to keep in mind when setting prices are:

- The *actual* price of the good or service (price perception)
- The prices of *related* goods or services (relative prices)
- The income of the buyers
- The preferences of the buyers
- The expectations of the buyers (What's a fair price?)

For example, if you're selling an expensive product, like an industrial machine, you might consider arranging finance so that the customer can pay back a relatively small monthly figure. If you can show that their monthly payback is greater than the amount they'll have to pay, you've almost guaranteed a sale.

If you sell small-value items, you could consider offering a pack of items at a discounted price. If you have some products lines that are not selling well, you could consider including a small number in a pack with the more desirable items. You could use the same principle when selling services. For example, a consultant could sell a "pack" of five consultancy days for a discounted rate.

Logic

We've seen that while logic plays a part in pricing, not everything is logical. That's why you need to consider *every* option carefully. For example, do you think you could charge a premium price for an item that's commonly available for free? Could you do so even when there's no difference between the free version and the one you're going to offer? It doesn't seem like a reasonable business proposition, but bottled water companies have been succeeding for years.

On the other hand, do you think you'd survive for long if you gave your product or service away for free? That seemed like a ridiculous business model until Google used it to become one of the biggest companies in the world. It's unlikely they'd have achieved the same level of success if they insisted on customers paying a monthly subscription for their services. There are many other examples like Angry Birds and the Unreal Engine.

Of course, those are software products, so there's no incremental production costs to worry about. Could the same model work with physical products? For example, could a company give away their jet engines for free on the proviso that they get the exclusive contract to maintain the engines for a defined period? Maybe not yet, but Rolls Royce have been offering a "power by the hour" programme for their maintenance services. Customers pay a fixed cost per operating hour. It allows them to plan their expenditure with certainty.

There are many other pricing models and, if you spend some time on it, you may be able to design a new one.

System 2

This chapter was included for two reasons. The first is because it's a very important area and every CEO needs to understand what's involved. The second is that it's another practical test of your System 2. Did you skip *any* of the details above? If you did, it's likely that you'll do the same when your staff are explaining things to you. The danger, of course, is that you'll accept misleading information.

Why would anyone give you misleading information? Some people might do it deliberately to influence you. Other people will do it because they're mistaken or have been misled. *You* can't afford to accept information without using your System 2 to analyse it and determine its validity. In other words, you can't "switch off" or skip over important information, even if numbers *are* involved!

The CEO's Job

It's likely that other people in your organization are primarily responsible for setting prices. You may even have an entire department dedicated to it. However, this is such an important topic that you can't abdicate responsibility. You must understand the pricing options available and their consequences. You must understand the impact of price changes, not just on the volume sold, but also on the bottom line. You *must* make sure that all pricing decisions are made using a System 2 decision-making process and not gut-feel. The success of your business depends on it.

FURTHER READING

Simon, Hermann, *Confessions of the Pricing Man: How Price Affects Everything*, Springer International Publishing.

15 PEOPLE

As CEO, you've got to deal directly with people and, in this chapter, we'll look at tools you can use to get better results. You've already seen the 4-1-1-4 tool in action. That should remind you that most things in life are non-linear. It should prevent you from accepting the notion that more is better because that's rarely true. When you're dealing with people, you need a similar tool. This one will remind you that people are different, and won't all react in the same way.

Of course, you *know* that already, but often, in the heat of the moment, that awareness can slip away. You implicitly assume that: "*They'll* get upset if we cut overtime", "*They'll* be delighted with a 10% bonus", "*They'll* react badly if we try to introduce change". Let's use a new tool to address that mistake.

The 1-8-1 Tool

The 1-8-1 tool is illustrated below. Let's use it to consider what could happen if you're going to introduce an organizational change.

 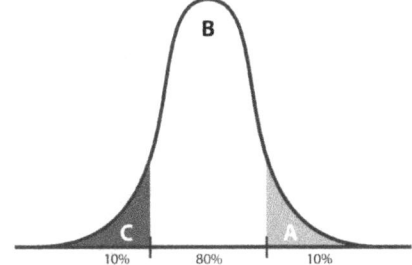

The graph shows the total workforce divided into three groups: on the right side of the graph – in Zone A – 10% of the workforce will be happy with the change. The 80% in Zone B will be moderately reluctant but can be persuaded. However, that 10% in Zone C will be highly resistant to change. They'll do everything they can to resist.

It's worth considering the people in Zone B in more detail because they're not a homogenous group. Those on the right of centre are positive-leaning but have reservations. Those on the left of centre are negative-leaning but are open to persuasion. If you do a good job of communicating the message and provide good reasons for the change, most of the 80% will go along. Just a point to remember: the figures quoted: 10%, 80% and 10% are only approximations, just like the 80%, and 20% of the Pareto principle. In a specific case, the 10% figure might be as low as 1% or as high as 50%.

So, what's the point of the tool? It should change the way you think about the workforce. It'll remind you that you're *not* dealing with a single group. There are at least three separate groups with different attitudes and requirements and you'll probably need three sets of tactics to bring them on board. You'll also need to monitor the people in Zone C very carefully because they could ruin your plans and your organization.

1-8-1 Tool Applications

The 1-8-1 tool isn't just about change. You can use it every time you're considering *any* decision that involves people in your organization. It applies, for example, in the following areas:

- Knowledge
- Productivity
- Honesty

Let's take *honesty* as an example. It's tempting to assume that everyone is honest in your organization but that's naïve. You shouldn't need to be reminded of the high rates of white-collar crime, much of which goes unreported. Let's see how the tool breaks that down:

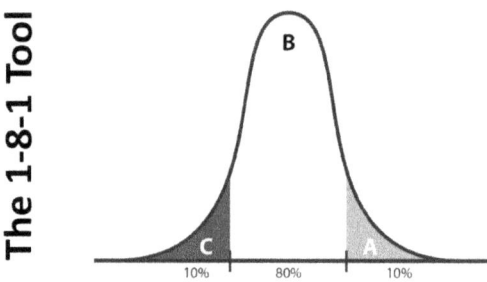

The people in Zone A are completely honest and will never be tempted to do anything even vaguely deceitful. The people in Zone B are reasonably honest and wouldn't do anything seriously wrong. However, the people in Zone 3 will take advantage if they get the chance. A small number are utterly unscrupulous. It's estimated that about 1% of the general population are *psychopaths*. That doesn't mean they're killers, but they *are* self-centred, dishonest and undependable. They'll even engage in irresponsible behaviour for no obvious reason. Incidentally, research suggests that the following professions have the most psychopaths:

- Number 3: Media workers
- Number 2: Lawyers
- Number 1: *CEOs!*

A 2011 Australian study found that 5.7 percent of white-collar managers could be classed as psychopaths! The 1-8-1 tool should prompt you to put systems in place that'll prevent this group from doing anything that could compromise your organization. However, it should also remind you that most people are honest and any measures you put in place should be proportionate.

Performance

Is your workforce hard working? The tool says:

- 10% are extremely hard working and need no supervision
- 80% are reasonably hard working but need encouragement
- 10% will do as little as possible and need supervision

What does that mean for *your* organizational design? Are you able to identify the people in Zone C or are they too well hidden?

Productive

Are your employees productive? That's not quite the same as the last question. The 1-8-1 tool says:

- 10% are extremely productive
- 80% are reasonably productive
- 10% have low levels of productivity

You might be tempted to assume that the 10% with low productivity are the same people that do very little, but that isn't necessarily the case. Of course, many people will be in Zone C in both cases, but you might find a hard worker (Performance – Zone A) who has low productivity (Productivity – Zone C) because they're working on the wrong things. That's often true of managers and CEOs!

The Halo Effect

The 1-8-1 tool highlights the mistaken assumptions behind the *halo effect*. The following diagram illustrates how your System 1 judges people versus the reality:

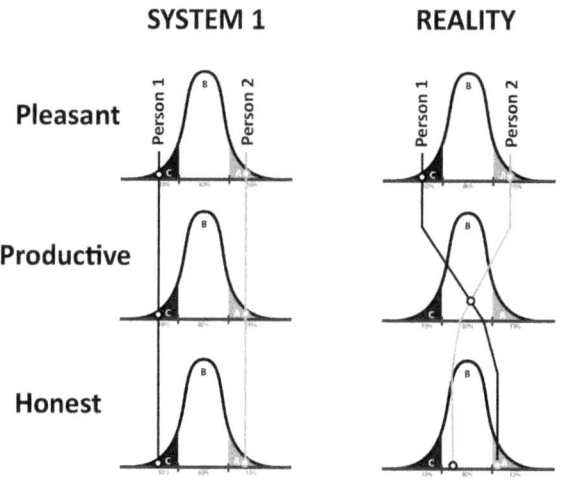

If someone is performing well in an area we can easily observe (for example, they're pleasant) then we *assume* they have other positive attributes (for example, they're productive and honest). If Person 1 is unpleasant you're more likely to believe they're also unproductive and dishonest. In fact, Person 2, who's your favourite, is less productive and honest than Person 1 – but you can't see it. This is another example of our "gut feel" decision-making process getting things wrong. Managers can develop an "irrational" dislike for a person and make their life difficult. The trigger might be something minor, like a casual remark or a glance. Suddenly, nothing that person can do is good enough and this results in unnecessary conflict.

Respect

You're responsible for creating a culture of *respect* for everyone in the organization. That means you need to listen carefully to everyone, no matter what their position or your impression of their capability. You've got to *engage* with them as if they were the chairman of the board – and if you promise them something, you must deliver. You also need to make sure that managers and team leaders are doing the same. If anyone tells you they're unhappy about how they're being treated, or you hear about anyone in that position, you need to investigate and establish the facts.

You're under the spotlight because you're the boss. If you disrespect anyone, for example by using nicknames or by making disparaging remarks, then you're signalling that this is acceptable behaviour and others will do the same. Some may even take it much further and the atmosphere could become toxic very quickly. As CEO, you're responsible for your own behaviour and for every manager and team leader in the organization. You can't afford favourites because that'll cloud your judgement, and you can't afford to create enemies either. Remember that 10% in Zone C who are vehemently opposed to your changes? It's tempting to regard them as traitors who need to be punished. Instead, you need to treat them with respect. You need to *understand* the reasons for their objections, accept that they feel threatened, and work to bring them along. Even if you find, eventually, that it's not possible, you need to treat them with respect as you face the next step.

Now some more questions for you:
- Are you good at delegating?
- Should managers usually make their own decisions?
- Should workers make decisions for themselves?

Write down your answer before reading on (Q15-1).

Your Feedback System

We're now going to discuss a communication technique that's very simple, but also very powerful. When you're dealing with equipment, you can use metrics to establish what's going on – you don't necessarily need to *see* the machines. When dealing with people, metrics are not enough – you need to *engage* with them.

**The solution is to get out of your office
once a day and *listen* to people.**

You can make the best use of your time by following these rules:

- Reserve the same time each day (e.g. 8.30am – 9.00am)
- Pick a random person each time
- Don't second-guess the person chosen
- Go to their place of work and engage with them
- Ask them for feedback and then listen
- Don't lead the conversation
- If you promise anything – follow up!

Now, let's look at each of these rules and the reason behind them.

Reserve the Same Time Each Day

This process should become a habit and that's only possible if it happens at the same time each day. It should also last for approximately the same duration. Block out the time in your diary and never schedule anything else for that time, no matter how tempting it may be.

Pick a Random Person

If you just walk around and talk to the first person who catches your eye, it's likely you'll only talk to people you like. That's human nature – you won't even realise you're doing it. You *must* use a system. Randomly select an employee number without knowing who it is, then look up the number and go and meet that person. That means you'll talk to a random selection of people in every area, from janitorial and security to utilities, production and finance. You'll meet some people who are negative and some who are positive. You'll meet a cross selection of workers, team leaders and managers.

Don't Second-Guess the Person Chosen

When you discover whom you've selected, you might be tempted to exclaim, "No! Not him!" and then select someone else. Don't do that; you could lose out on valuable feedback, no matter how uncomfortable it might be for a few minutes.

Go to Their Place of Work and Engage With Them

Engage with people in their own environment. They're more comfortable there and you'll also get to see their working conditions. If the area is particularly noisy, you may have to find a quiet place close to their work. Ideally, you don't want other people listening when the two of you chat because that could inhibit what they say. As a matter of principle, you should inform the team leader before talking to someone who reports to her.

Ask Them for Feedback and Then Listen

It's tempting to try to direct the conversion into areas *you're* interested in, but it's more useful to find out what the person is worried about. Where possible, let them lead the conversation. Have a few non-directive phrases available to keep the conversation going. You might use some from the following list:

- How's the job going?
- Why do you say that?
- What's the reason for that?
- Tell me more about that.

You should try to avoid conversations involving criticism of their direct supervisor. If they bring it up, ask them for a recent example of the behaviour they're complaining about, then move on to other areas. Don't get trapped into defending the actions of their supervisor, even if you think they were justified. If the allegations are serious, you need to follow up with the manager and team leader involved.

Don't Lead the Conversation

You may need to read between the lines to find out what's going on, and it might take ten or fifteen conversations before you get a sense of what's *really* happening. You might be tempted to ask leading questions to speed things up and confirm your suspicions. Be careful, because you could drive the conversation in the wrong direction and make things worse. Let's say you ask: "What problems are you having with the new equipment?" The person may not have noticed any problems before you asked. Now, however, he may give you a list of things that could be improved. What's worse, he could start looking for problems and his dissatisfaction will grow whenever he finds a minor issue. It's better to ask the more general: "Do you have any challenges in your job"?

Follow Up!

Take a notebook with you when talking to people. It's useful for keeping a record of anything that might be important (and they'll be flattered that you regard it as noteworthy). The notebook is *essential* for making a note of anything you promise. Write it down, then follow up later – it's not enough to ask someone else to do what you promised. For example, you might promise someone a new chair because the old one is broken. You ask their manager to arrange it – but don't just forget about it. Follow up a week later to make sure they received it. That's important because your credibility is on the line.

Objections

Maybe you don't think it's worthwhile walking around and meeting people in your organization? The alternative is to sit in your office and have feedback filtered through team leaders and managers. Filtered feedback won't give you a comprehensive picture of what's *really* going on. Managers will give you a version that'll be coloured by their own fears, expectations and desires. Some will lie ("everything's fine in my department"), while others will attempt to tell the truth but won't have all the facts and tell you stories instead – and that'll be dangerous for decision-making. There are many good reasons for meeting people, but perhaps you still have reservations?

Let's address a few of them:

- They're too far away
- It's disruptive
- I haven't got the time

They're Too Far Away

If the people in your organization are spread around the globe, then it's going to be difficult to visit a random person every morning. Instead, when you're visiting each remote location, put aside a day to meet people.

Follow the same rules but leave yourself fifteen minutes between conversations to digest what's been discussed and take more notes. That's necessary because you're getting a lot of information in a short time and you'll forget much of it if you don't record it immediately afterwards.

It's Disruptive

It doesn't have to be. First, let people know that what's going on. Talk to the managers and team leaders. Explain why it's important for you to listen to people at every level in the organization. Listen to any feedback – particularly about any dangers they might anticipate. Do what's necessary to minimize those dangers but insist on talking directly to people. If someone is adamant that you shouldn't do it, you might need to question their motives.

I Haven't Got the Time

If you haven't got time, it's likely you're doing the wrong things and you're not being effective. This task feeds your understanding of the organization and how it works. It reinforces the control systems that are in place and it ensures that new systems are given a fair chance. It allows you to *speak with authority to your managers,* because you're not depending on them for all organizational information. And it also helps you promote the culture you want and make sure it's growing throughout the organization. You may also get great ideas for improvement from people whose voice is rarely heard. These discussions may well become the most important and visible aspects of your leadership.

Managing Biases

As you read this page, you're seeing it *through* your eyes. However, you're not *seeing* your eyes. That might seem obvious, but it has ramifications. Before we discuss them, try this little experiment (it's worthwhile!)

- Close your left eye
- Fix your right eye on a spot (e.g. a small picture) on a plain wall about five feet in front of you
- Don't move your eye
- Make a fist with your right hand and then stick your thumb up
- Hold your hand at arm's length with the thumb upwards
- Move it around until you notice the top of the thumb has disappeared!

That's right: each of your eyes has a blind spot. It's caused by the optic nerve passing through the optic disc. That's a physical limitation of the eye. What's interesting is that you normally can't see this blind spot because the brain fills in the blanks. Your brain is lying to you about the world at a fundamental level.

When the managers, team leaders and workers in your organization look at the world, they're seeing it through their biases, but they can't see their biases either. However, if *you* watch and listen carefully, you'll get to recognize them.

We've seen that people suffer from the *"confirmation bias"*, the *"self-serving bias"*, the *"overconfidence effect"* and the *"action bias"*, to name just a few. These biases mean that people will do illogical things. They'll react in illogical ways. Part of your job is to educate your team and try to minimize those biases. However, that's not enough. You'll also need to build a *system* that'll compensate for those biases.

Just to stress that point. You *know* those biases exist (if you've got any doubts, you should do more research to confirm their existence). You *know* that people don't always make logical decisions because of these biases. If you do nothing, you're *accepting* the near certainty that people on your staff will make serious mistakes. That's another example where *not* taking a decision is the same as making a bad one. If you do nothing, *you're making the decision to accept the risk.*

Decision-Making Styles

We looked at decision-making styles in an earlier chapter. You probably remember the five styles:

- Charismatics
- Thinkers
- Sceptics
- Followers
- Controllers

As we've seen, each of these styles has advantages and disadvantages. Let's examine how things might work out when people with these styles work as managers or team leaders.

Charismatics

Charismatics are positive and pro-active. They work well on short duration projects. However, they can struggle on longer projects because they get bored and want to try something new. They can be inspirational managers but may not be good at administration, often finding it difficult to keep paperwork up to date.

Thinkers

Thinkers are great at analysis and enjoy working on business models. They'll personally crunch numbers and work through the details of any proposal. They're also good managers, giving credit where it's due, but may find it difficult relating to their staff and can become very impatient if they detect sub-standard work.

Sceptics

Sceptics can be abrasive, both to their own staff and their colleagues. They can also be aloof, so people can find them difficult to work for. However, once they have a goal, they'll let nothing stand in their way. That's often viewed as a good thing – but it can have negative consequences.

Followers

Followers can make good managers, particularly if you want to maintain the status quo. They relate well to colleagues and to the people who work for them and are usually regarded as good, solid managers. However, it can be difficult to get them to change. They'll resist in a passive way – simply by not doing things.

Controllers

Controllers are *always* looking for threats. They don't trust anyone and resist delegating important tasks. They don't believe in consensus and will try to decide everything themselves. They can be very difficult to work for because they want every trivial detail to be perfect. They'll also take all the credit for everyone's ideas and blame others for their own mistakes. This behaviour tends to undermine morale. However, they can be very hard workers.

As you can see, there's no such thing as a perfect type, just as there's no such thing as a perfect person. The best solution is a mix of the best characteristics from each type and that's why working together is so important. Individually, we are prone to mistakes but collectively, we can surmount any obstacle. As CEO, you must work with each of your managers to maximize their positive attributes and minimize their negative ones. You must also make sure that they're doing the same with the people who report to them.

Business Process

Hiring is an example of a *critical business process* in your organization, but do you actually believe that? Would you agree or disagree with the following statements?

- People are one of the most important assets in your organization.
- It's very important to select the right people.
- Selecting even one wrong person can have significant consequences.

Write down your answer before reading on (Q15-2).

Most CEOs would agree with those statements. Now, how about these ones?

- We select most of our people using an interview.
- The final hiring decision is left to the manager.
- We don't have a feedback loop for our hiring process.

Write down your answer before reading on (Q15-3).

If you answered "yes" to one or more of the last three statements, then you have a sub-optimal process and it's likely that you've been hiring the wrong people. In fact, you may be making two errors. First, you could be hiring people who are unsuited for the job. They may not have the technical skill or the right attitude. They might be resistant to change or disruptive, so you're *failing* to exclude people who shouldn't get in. Second, you may be *excluding* brilliant people because of illogical biases. You may be missing out on people who could make an outstanding contribution

Selection

Most managers agree that having the right people is critical for success in any organization. That means every organization should take extraordinary care to select the right people. So, let's look at *your* system. The last time you hired someone for a management or supervisory role, what was the *primary* selection method used? Select one item from the following list:

- Telephone interview
- Structured questionnaire
- General mental ability test
- Unstructured interview
- Simulated work test
- Technical test
- Personality test
- Reference checks

Write down your answer before reading on (Q15-4).

When you use a selection process, you're trying to forecast a person's future work performance. The trouble is that most people have a go-to solution that's close to useless. Did you pick any of the *interview* options as your solution? Many studies have shown that interviews are *not* good predictors of performance. They simply don't work.

The costs of making a mistake are significant. The US Department of Labour and Statistics estimated that the average cost of a bad hire is up to 30% of their first year's salary. And the problem is widespread. In one survey, 66% of employers said they experienced the negative effects of bad hires. Of these, 37% said the hires reduced employee morale, another 18% said they soured client relationships and 10% said they caused a decrease in sales!

Why should selecting the right person be *so* difficult? The answer shouldn't come as any surprise. It's because System 1 is in charge and it's faced with a tough question: "How will this person perform in the future?" It doesn't know the answer, so it substitutes another question, an easier one: "Do I like this person?" Unfortunately, the right answer to *that* question is unlikely to be the same as the first. The discrepancy is caused by biases that every interviewer suffers from. These include:

- *The Halo Effect:* This candidate is pleasant and cheerful; she'll be a great worker.
- *The Horns Effect:* He has a weak handshake; he'll never be able to deal with customers.
- *The Similarity Effect:* He came from a small town like me; he must have a terrific work ethic.
- *Appraiser Bias:* She supports my political party; she must be intelligent.
- *The Primacy Effect:* Her smile and warm handshake set the tone for our interview.
- *The Contrast Effect:* The last candidate was a disaster, but this one's brilliant.
- *Leniency/Strictness Bias:* Mary's average assessment is much lower than mine; she's too strict.

The Gateway

The hiring process is the gateway into your company. If you leave it to chance, then the people who come through won't be the brightest and the best – at least, not by design. So, *why* is one of your most important business processes in its current condition? Perhaps you haven't given it much thought? Perhaps you weren't aware that there was a better way? Perhaps you just accepted the status quo? It's important to dig into those reasons because it's likely that the same forces are at work in other areas.

The Process

What should the hiring process look like? The following is a brief overview:

- Draw up a job specification (what will they do?)
- Identify and list the skills required for the job.
- Identify and list the personal characteristics required.
- Review and eliminate any unnecessary requirements.
- Devise a set of tests for each of the requirements.
- Include a test for a fixed or growth mindset.
- Devise a questionnaire to evaluate experience.
- Develop scoring and accept/reject criteria.
- Contact the prospects and carry out pre-screening.
- Carry out preliminary tests – score the results.
- Carry out the structured interviews.
- Score the results.
- Hire the candidate with the highest score.
- Review performance at three, six and nine months.
- Review performance again at one, two and three years.
- Modify the process based on performance reviews.

Let's review a few of the key points. First, everything you do in this process must be made *explicit*. For example, if you want to someone with an outgoing personality for a job, you must break it down: what does an outgoing personality *mean*? You might specify that the person smiles, has a firm handshake and makes eye contact for at least two seconds when you meet.

When you carry out the first review, you should make sure that the list includes *only* the essential requirements. For example, should the person have a degree? Qualifications are usually just a matter of convenience for the recruiter; it's rare that they're necessary. So, unless a degree is *absolutely* required, leave it out. That way you're not unnecessarily eliminating good candidates.

A key requirement is that the recruit should have a *growth mindset*. This means they believe that hard work will improve their knowledge and performance. This contrasts with people who have a *fixed mindset* and think they were born with all the innate talents they need. This belief can lead to defensive behaviour and a lack of effort when things get tough.

A small group of people, including the prospective new hire's manager, should evaluate a list of requirements. They should also be involved in the development of the tests, the questionnaire and the scoring. The group should use the decision matrix and the decision-making process to arrive at their conclusions.

A structured interview can be carried out using at least three interviewers and a prepared list of questions. No other questions are asked, except to clarify the response to a question. Each interviewer notes down the answers given by the prospect and these answers are scored immediately afterwards, using prepared criteria. The scoring is done individually and then the three interviewers compare results. If there is a big discrepancy between scores for any question, the interviewers discuss the reason for their differences and try to achieve consensus. In any case, the average of all three scores is used. The records of each interview are filed and are later used in the evaluation process.

Whenever possible, the team will *test* the person on their capabilities. For example, if a candidate says they've been working on a process, the team will ask detailed technical questions that'll confirm his knowledge of that process. Those questions will be prepared before the interview with sample answers.

The prospective manager is involved in all aspects of the hiring process, but *she doesn't have the deciding vote*. She can't veto a candidate or select a low scorer. That's to prevent her System 1 biases from wiping out the benefits of the selection process.

Feedback Loop

The last three steps in the process are part of the feedback loop. Is the process working? Can it be improved? The only way to tell is by checking results. If the results are good, it probably means that the process is working – although improvements might still be possible. However, if new hires are not working out as expected, then the process *must* be improved. The records from the testing and structured interviews are available, so a new team can investigate what happened and determine how the process can be modified.

Implementation

The process above is logical and straightforward to implement. You can see how it'll reduce the effects of bias on the final selection and, consequently, why you'll get better people in your organization. Do you have any reservations about using it? If you do, it's another demonstration of System 1 thinking. Yes, it's a lot easier to interview someone and then make up your mind using gut instinct. However, it's also likely that you'll spend more time dealing with the fallout from wrong choices. Are you still tempted to spare a few hours in the short term and risk wasting weeks or months in the medium and long term?

Induction

The first three months are the most important period in a person's tenure in any job. This period sets the person's beliefs about the culture in the organization and its expectations. It also builds or recalibrates a person's "*social scripts*". We've seen that social scripts are a series of behaviours and responses that occur under specific circumstances. Two researchers, Ellen Langer and Robert Abelson, found that different scripts could be triggered in the same situation using different patterns of words. Once triggered, the person will follow a series of steps automatically, even when they make little sense (System 1 in action). A new hire is likely to import some scripts from a previous employment, but she'll also build new ones during the first few months on the job.

New people are often paired with experienced ones during the first three months. Sometimes, the experienced person is technically competent *but not the best person to provide a good example* – and that can be a serious error. The new hire may learn a set of social scripts that are inappropriate for the organization. It can be very difficult to change those scripts and expectations at a later stage when they are deeply embedded. It's worth making the effort to get it right at the start.

Firing

There will be times when you must consider getting rid of someone. First, make sure that you've followed all legal requirements. You should consult a solicitor or lawyer and ensure you're fully compliant – no shortcuts or fudges – because otherwise it will be a very expensive experience. Next, put some distance between yourself and the situation because it's possible that emotions are clouding your judgement. Start with the facts. What's happened? Who did what? For example, how would you evaluate the following statements:

1. He looked like he was going to kill me.
2. He pushed me out of the office.
3. She walked past and ignored me.
4. She didn't respond when I asked her a question.
5. She favours other people in the department.
6. He screamed at me for several minutes.

Write down your answer before reading on (Q15-5).

You need to separate out facts from assumptions. If everyone is telling the truth, statements 2, 4 and 6 are facts. If other people were present it should be possible to *confirm* that these incidents took place. On the other hand, statements 1, 3, and 5 are *assumptions*. For example, in statement 1 the person may have had a strange look on his face, but we can't assume we know the reason. He could have been suffering from heartburn! Likewise, in statement 3, the person may have walked past, but may have been preoccupied and never noticed them.

You also need to avoid attributing *reasons* for a person's behaviour: "He gave the job to the other candidate *because he didn't like me.*" "She gives out work assignments early *to stop me getting the easy ones.*" These statements combine facts with assumptions. Strip away those assumptions and work with the facts.

Of course, you can't assume that everyone is telling the truth either. Treat all allegations very carefully. We tend to believe people in our circle, particularly if they seem agreeable. However, that bias could have negative consequences. Suppose a manager complains to you that a worker is causing problems and wants to fire her. You concur because you take his report at face value. But what happens if he's lying or distorting the facts? You've fired an innocent person and she'll probably sue for wrongful dismissal. What's worse, people in your organization will see that you support managers even when they're wrong. You've just promoted a "them versus us" culture. So, check all the facts and get the full story from both sides. Everyone *could* be telling the truth, but be aware that some people will "bend" it when under pressure. When different stories are being told, don't assume that the person you prefer is telling the truth and don't depend on a single unconfirmed source; dig deeper.

Even when a person is behaving badly, it can be difficult to get rid of them. It's important to have a well-crafted contract of employment for occasions when that becomes necessary. Again, consult a solicitor or lawyer to make sure you have the best advice incorporated in the document. It can also be useful to have someone in your organization monitor any proposed changes to labour law, so you can have input via your industry group.

Working With People

As we've seen (with the 1-8-1 tool) most people want to work hard and do a good job. Sometimes the organization (i.e. you) puts obstacles in the way of that happening. Here are some of the common reasons why people can't do their job properly:

- Micromanagement
- Insufficient training
- Rapid uncoordinated changes
- Unclear instructions
- Contradictory instructions

Micromanagement

If a team leader or manager feels the need to specify and check every aspect of the work carried out by an employee, then they are micro-managing that person. It's very disruptive and disheartening for the employee and a waste of time for the team leader or manager. One decision-making type is particularly prone to this type of behaviour (can you remember which one?). It can be hard to change this, so counselling may be require for the person concerned.

Insufficient Training

When people aren't trained properly, they won't be able to do their job properly. This area is so important that it warrants its own chapter, so we'll come back to it.

Rapid Uncoordinated Changes

This can happen because nobody is sure exactly what going on. For example: there was a huge emphasis on getting more units out last week. However, yesterday, the most important thing was to get the quality right and extra resources were dedicated to inspection. Today, costs are too high, and all overtime has been cancelled. The problem can be traced back to the CEO (you) because priorities are either not clear to people or are being changed on a whim. It's likely that you're not communicating things clearly.

Unclear Instructions

This can take several forms. It could relate to the work instruction for a job. There *should* be a written procedure in place, but it may be missing, incomplete or badly written. If that's the case, it's likely that the job will not be done properly. It could also relate to the verbal instructions given by a team leader or manager. They might ask for something to be done but phrase it so badly that the intent is not clear. This can lead to confusion and frustration. The solution is to train those in supervisory roles so they understand how to give crystal clear instructions.

Contradictory Instructions

This can happen when a team leader or manager gives an instruction that conflicts with a written procedure or with an instruction from someone else. Again, this causes confusion and frustration and can only be avoided if everyone is aware of the base procedures and there are clear lines of accountability. For example, managers shouldn't override the authority of team leaders.

The Unpack Tool

Communication is very important in an organization and it often doesn't get the attention it deserves. Some people are great communicators, but many are not. The *unpack tool* is useful when you're on the receiving end of people who have problems in this area. Consider this:

"The consultant said that sales have gone up 10% and costs are down by 5% but the share price hasn't moved, so now's a good time to buy."

- "So, let me unpack that."
- "OK."
- "Sales have gone up by 10%? Do we have confirmation of that?"
- "Actually no, not yet..."
- "OK, and costs have gone down by 5%? What were they before?"
- "I'm not sure."
- "And the share price hasn't moved? Has the analyst been talking to the company about the fundamentals?"
- "I don't know; I assume so."
- "OK, let's hold off until we get more information."
- "Right!"

The unpack tool gives you a chance to pause and consider the evidence. It gives you permission to take each point in isolation and check its validity. You can then put it all back together and see if it still makes sense. Here's another example:

"Can we get John to chase something? Mary and her team are working on the Forrest job and the Johnson job never got here – you know we were expecting it? Of course, the new equipment has arrived and Jack is working on that and May is out for the rest of the week so that's a problem. Mary is afraid he'll get sucked in if he starts because they have another job coming this evening. What do you think?" This time you can use the tool like this:

- "So, let me unpack that."
- "OK."
- "John works for Mary, right?"
- "Yeah."
- "And the problem is that the Johnson job never arrived?"
- "That's right."
- "And May is out for the week?"
- "Yeah."
- "And Jack is also tied up?"
- "Yeah."
- "So, John is the only one available?"
- "That's right."
- "And Mary is afraid to release him in case he's not back for the Forrest job?"
- "No, for a new job that's coming in tonight."
- "Ok, just ask her to release him until 6.00 this evening."

In this case, you're taking a message that has many elements and you're breaking them into easily digestible chunks. You get confirmation about what you've understood correctly, and you get immediate feedback about anything that you haven't. The unpack tool is always useful when you're being briefed. It allows you to stop the flow of information at any time and confirm that you understand every element. It prevents information overload.

Those features make it useful for every person in your organization. It significantly reduces the chances of miscommunication and could make a difference to operational effectiveness. Shouldn't you be actively promoting its use in your organization?

Final Point

Just one additional point: if people are coming to you with problems like John's availability, *you're not delegating enough* – you may be micromanaging your staff. You need to let go – put in a robust system and then trust it to deliver.

FURTHER READING

Dweck, Carol, *Mindset: Changing the Way You Think to Fulfil Your Potential*, Little, Brown Book Group.

16   CHANGE

Imagine that the location of the controls in all new cars is about to change. There'll be no more automatics, only manuals, and the pedal positions will be swapped: the accelerator will move to the centre, the clutch to the right and the brake to the left. The gears will also be changed: first, second and third will be forward and fourth and fifth will be back. The reason is that scientists have determined that the new layout is safer. What's your immediate reaction? Would you be tempted to rush out and buy a car *before* the changes come into effect? Would you buy second-hand cars for a few years rather than new ones? Change is hard, isn't it?

The Status Quo

Why do people resist change? One reason is that System 2 is lazy: it doesn't want to do any work if it can help it. However, when you make a significant change, System 1 can't work properly anymore. Today, you can drive effortlessly because your System 1 is doing all the work. You don't have to think about changing gears or which leg is pressing which pedal. When the change comes, your System 2 will have to think about *every* action and that's hard work, it'll take a lot of energy. What's more, you'll inevitably make mistakes and that'll cause stress and discomfort. You realise it could take many months before you'll find driving relaxing again.

There's also something else, a nagging feeling at the back of your mind. *Maybe you won't be able to do it.* Maybe you'll never recover the proficiency you built up over your years of driving. *You're afraid.* It's no wonder that you want to stick to the status quo. It's no wonder that you're considering buying two cars, so you can put off the evil day for as long as possible.

Reasons to Resist Change

There can be many reasons to resist change, some rational and some totally irrational. The irrational ones are driven by fear, confusion, hatred, envy and jealousy. They represent System 1 at its worst. It's unlikely you'll hear about them directly, but you might get glimpses as people lose their temper and say more than they intended.

You'll hear many rational reasons why change won't work and you'll need to carefully evaluate these. If there's a sound basis for the objection, you need to acknowledge and determine the best way to handle it. However, many *apparently* rational objections are driven by emotion. System 2 has been told to find a way to stop the changes and it looks for anything that could be taken as a rational argument. Many don't stand up to scrutiny, but they can slow things down and even become a pivot for revolt unless handled carefully.

The Stages of Change

Change can have the same effect on people as a death in the family. The Kubler-Ross model is often used to illustrate the stages that people go through under those conditions. The diagram below is a variation of that model. The vertical axis tracks *morale* and *performance* because they're linked. The horizontal axis is *time.*.

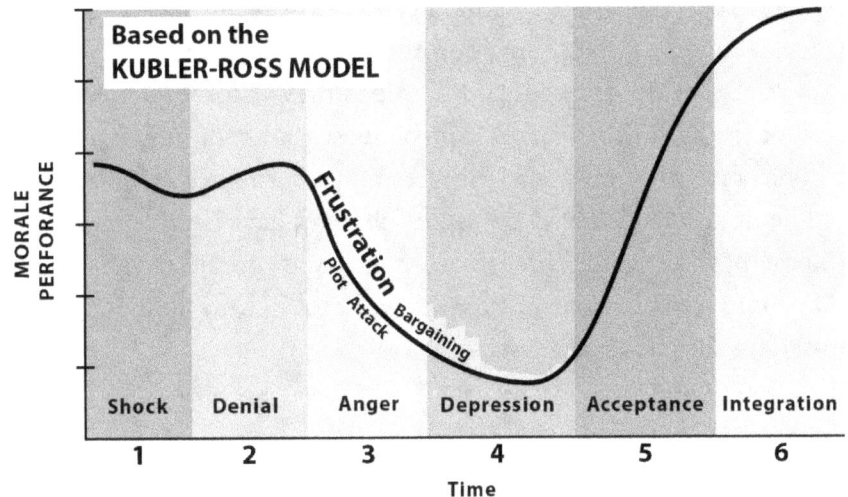

The first stage is *Shock*. People are astonished that change is about to happen. Morale and performance can drop as they begin to recognize all the things they'll lose. That could include the competencies they've built up over the years. It might also include their circle of friends, their network of relationships and their social position in the group.

The second stage is *Denial*. They don't believe it's going to happen. It's all a mistake. Management will back down at the last moment. The board will stop it from happening. Morale and performance may increase slightly as people reassure themselves that it's all going away and nothing is going to change.

The third stage is *Anger*. It's going ahead, so now they're looking for people to blame. Don't the managers realise that it's a disaster waiting to happen? They haven't thought it through. As changes continue, frustration kicks in. Problems arise that never occurred before, and it's all due to the new practices. Some people are quietly plotting ways to sink the project. They're looking for a scapegoat to blame. Others are openly attacking anyone associated with the change. Some people want to *negotiate* – they're willing to implement *some* of the developments if the rest are dropped. Morale and performance are dropping.

The fourth stage is *Depression*. The changes are still going ahead and people are depressed. There's still some smouldering anger and frustration around, but both morale and performance are at their lowest.

The fifth stage is *Acceptance*. People are beginning to realise, at last, that there may be some good things associated with the change. They're also beginning to let go of things they've lost and are starting to work through the changes. Maybe it isn't all bad; in fact, it might even be better in some ways. Morale and performance are moving up.

The final stage is *Integration*. The changes have been accepted and are now part of normal life. Performance and morale have increased and people are able to hand over the work to their System 1s again. A new status quo has been installed.

Why?

You can see that change is a difficult business; it can affect everyone, not just those who are directly involved. So you might ask: *Why? Why not leave things as they are? Everyone would be much happier.*

The simple answer is: Because it's your job. You've been hired to make a difference and that means making changes. It's true that if you don't make any changes people will be happier – but only in the short term. In the medium term, they won't have a job because your competitors won't be standing still – and you'll be responsible. You'll have traded an easy life in the short term for people's future.

The 1-8-1 View

As usual, where people are involved we can use the 1-8-1 tool. This time, we'll use it to classify people's reactions to *significant* change, and you can see, there's a slight variation:

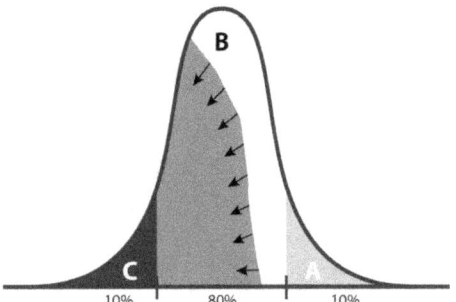

- *Zone A:* Represents the 10% of staff who are happy to make the changes. The negative aspects of the Kubler-Ross model don't really apply to this group.
- *Zone C:* Represents those people who are *absolutely* against change and will fight tooth-and-nail to prevent it from happening. They'll plot, attack and subvert everything and everybody to get their way.
- *Zone B:* Usually there's a smooth transition between Zone A and Zone C with most people in the centre. However, the corrosive effects of the opposition can pull people in a negative direction.

The questions you need to answer are:

- How do I harness the positive attitude of people in Zone A?
- How do I move people in Zone B away from Zone C?
- How do I identify the people in Zone C and counteract their negativity without driving them underground?

You should build the answers to those questions into your plan.

Steps to Change

How do you ensure that your changes will be successful? There are no guarantees. However, there are several steps that, if *not* followed, will almost certainly guarantee failure. The steps are as follows:

1. Make sure you're *certain* about the changes
2. Create a sense of *urgency*
3. Establish an implementation team
4. Benchmark your *current* operation
5. Create a *vision*
6. Explain the *reality* of change to stakeholders
7. Communicate the vision
8. Empower people to act
9. Make sure your unwavering support is visible
10. Create short-term wins
11. Consolidate improvements and build on them

1. Be Certain

We've seen that change is difficult and uncomfortable. It will affect *you*. There will be times when you'll be frustrated by the lack of progress. You'll be shocked when friends begin to attack your ideas. You might even find yourself under personal attack. You'll be tempted to divert resources away from the project. If you're not certain about what needs to be done, and why, then the project is doomed before it starts. You're going to need a good reason to keep going during the dark times. You'll need a strong reason to overcome the challenges you'll face. But what if you *do* give up? Then you'll have lost all the progress that was made. More importantly, you'll have lost credibility. People who supported you won't trust you as much the next time. People who opposed you will be encouraged to confront you again. It'll be much harder to get even small changes implemented.

The greatest danger will come several years into the project. Progress has been slower than you expected, and you hear about a different approach that promises better results. Immediately, you will be tempted to drop the current project and start a new one. That'll please your detractors, but it betrays your supporters and makes it less likely that the next project will succeed. That why you need to be *certain* before you start.

But how can you be *certain* that the decision is right? Get a small team of trusted colleagues to help you and use the tools in the decision-making chapter. Make sure that the trigger you've identified is valid and then review as many options as you can. Use the Decision Matrix to select the best option. Once it's been selected, stand back and give yourself time to think about it. Ask your colleagues to look for weaknesses and criticise it. When you're happy that you've considered all angles (particularly the negative ones) and you're willing to defend it no matter what happens, then the decision has been made. Record the decision with all the reasons and justifications (to remind yourself later) and then stick with it.

2. Urgency

The question you'll be asked throughout the project is: "*Why should we do this?*" It's a reasonable question and you need a convincing answer, one that'll drive action. It should have a *logical* element and an *emotional* one.

There are two ways to approach it. You could give a negative reason: "We'll lose out if you don't do this." Alternatively, you could provide a positive and uplifting reason: "We'll all gain from this." There should also be a time constraint: "We have an opportunity to become the biggest and best company in the sector but we need to complete all the changes within the next two years."

There are some dangers. Let's consider a common reason that people use to drive change: "The company will be forced to shut down unless we make these changes." If that's credible (and it must be) then it appears to meet our requirements. There's a logical element: "We want to keep working to feed our families", and there's also an emotional one: "It would be devastating if the company goes under. I might never get another job." It's likely that most people will be convinced and willing to make the change. However, can you think of any argument for *not* using this argument, in its present form?

Write down your answer before reading on (Q16-1).

One concern is that it might make one group resist the change even harder. Ask yourself: who'd gain if the project failed? In this case, the answer is people who're close to retirement age and who'd get a big redundancy settlement on top of their pension if the company went under. That's an example of a message having the reverse effect to that intended. It could create a strong resistance that'll secretly sabotage your efforts, and it's likely that they'd have the knowledge and the connections to succeed. Make sure you're not motivating *anyone* to oppose your changes. Being too negative might also prompt some of your best people to look for a job outside the organization.

3. The Implementation Team

The next step is to set up an implementation team. The key members must be the line managers and team leaders of the areas directly affected by the changes. That's important because otherwise there'll be a conflict between people who are trying to implement change and the people who are trying to carry out the day-to-day activities.

The team must have the resources to do their job properly and the most important resource is *time*. Some CEOs expect people to work on projects while still doing their "day job" but that doesn't work – the "day job" always gets the upper hand. "Urgent" trumps "important". People on the team need *dedicated* project time, during which they're not available for any other reason, no matter how urgent. That means someone will need to take over some or all their normal duties for the duration of the project.

The team members must be *trained* to carry out their task. The training will include how to manage the project, how to develop new ideas, how to avoid groupthink, how to make decisions and much more. Many of the concepts described in this book will be relevant to their work.

The team must have the *authority* to make changes. That doesn't mean they can do what they like. You need to provide clear guidelines as to what's acceptable and what's not. For example, it should be acceptable for them to make changes to the process, but not to make changes that could damage the quality or integrity of the product. The guidelines need to be broad enough to give them flexibility while still making sure that the core attributes of the business are maintained.

4. Benchmark Your Operation

Imagine introducing new procedures intended to improve productivity. Most people don't like the changes (as expected) and managers are telling you that they're making things worse, not better. How do you tackle *that* problem?

You must *benchmark* the operation *before* you start. This provides a *reference point*. You'll know what the operation was like using the old procedures. Later, you can compare the results before and after the project to determine what was achieved.

However, it's likely that your existing metrics won't be good enough for benchmarking. They won't have the *granularity* you need to pinpoint the problems you're going to face. For example, suppose you're measuring the number of sales calls made each day. After the team has rolled out the changes, you can see that calls are down. But why? Is it because of the new procedures? Is it because some people are deliberately working slower? Is it because they aren't using the new software system properly? The overall figure won't tell you, so you need more data to answer questions like:

- Are everyone's figures down, or only some people's?
- What's the difference between highest and lowest performers?
- What tasks are taking the most time?
- What tasks have increased in duration?
- Have tasks been added?
- Are old procedures still being used?

The team should identify likely questions at the *start* of the project. That way they can put measurement systems in place and get a set of benchmark results so they'll be able to pinpoint the reasons for any performance changes – good or bad.

5. Create a Vision

What will your organization look like after this project? People need a *vision*. They need to have something to strive for. And don't forget to identify what's in it for them. The team should be involved in the development of the vision. They must be brought in so they can explain it to others and defend it when things get tough.

Even if things look black and you're struggling to avoid bankruptcy, you need to give people a positive vision to keep them motivated. For example: "When we introduce these changes we'll have a solid company again. We'll be back on a five-day week and people will be able to go on their overdue holidays." Of course, the vision must be based on reality.

6. The Reality of Change

On 13th May 1940, Winston Churchill gave his first speech as wartime prime minister. The Nazis had occupied Poland and were now attacking French and UK forces. He needed to raise morale so that people wouldn't give up. However, he knew better than to raise unrealistic expectations. He held out the prospect of eventual victory, but the core message was sobering:

> *"I have nothing to offer but blood, toil, tears and sweat. We have before us an ordeal of the most grievous kind. We have before us many, many long months of struggle and of suffering."*

That speech was a masterpiece. Not long afterwards, the British and French armies were smashed, and on June 17th, France surrendered. Imagine if he'd insisted that things were going to be easy? Suppose he said that they'd have a quick victory? His credibility would have been destroyed. Instead, people appreciated the honesty and trusted him to lead them for the next five years.

You're in a similar position at the start of a change project. Of course, you're optimistic. The vision has been shared with everyone, you know what must be done and you're feeling confident. What could go wrong? Now is the time to accept that things *will* go wrong. Now is the time to explain the *reality* of change to all the stakeholders.

You need to explain to the board that there *will* be a dip in revenue but that's necessary to achieve the next level of performance. Of course, you'll do everything you can to minimise the dip and keep it as short as possible, but it's likely to last a significant time. At this point, they'll ask you for figures, how much of a dip and for how long? You should avoid answering those questions (like Churchill did) because if you give numbers you'll be *wrong* – you'll be too optimistic. It's going to have a larger negative effect than you're willing to accept.

You need to prepare the managers and team leaders for what lies ahead. They need to understand that things are *not* going to go according to plan. There will be problems. Tempers will flare. People will get frustrated. They *must* be ready to handle conflict. You need them to support the project when things go wrong so they must understand what's likely to happen and be prepared for it.

You also need to explain the reality of change to workers and staff. They'll need to prepare for discomfort and loss. Many of the things they've grown to rely on are going to be different. It's going to take extra effort to do their job for a while. Of course, there's a good reason for the change and a vision that'll keep them going during the difficult times, but they shouldn't underestimate the challenge.

7. Communicate the Vision

You need to communicate the vision to everyone and you don't do that once or twice, or even three times. You do it *every time* you communicate with people. You do it at meetings, at performance reviews, at communication sessions and Q&As. And you don't stop when things get tough. That's the time when people need it most.

It's important that the vision comes *from you*. This is something you can't delegate to others. It's also critical that your actions match your words. If you tell everyone how important that project is and then divert resources away from it, people will read between the lines. You must also make sure that the *same message* is coming from all the managers and all the team leaders. Sit in on some of their meetings and make sure they're promoting the vision, and make sure it's the same one you're promoting!

8. Empower People

Make it easier for people to make changes. If there are obstacles, then *you* must make sure they can be removed. For example, there might be a system that makes change difficult. There might be a manager or team leader who is preventing change from taking place. You need to identify those blocks and do whatever's necessary to overcome them.

Sometimes the problem is an internal one – people can't *imagine* any alternative to what they've been doing for years. *It's always worked, so why change it?* It may be necessary to give people training in innovation, so they'll be open to creating new ideas.

While it's important to remove the obstacles to change, you should never throw away all constraints. The changes must be implemented in a controlled way – the alternative is chaos. For example, suppose you allowed people to make any change they like. Three months later you find that customers are complaining about getting defective products or incorrect advice to solve their problems. You've damaged the organization's reputation and you'll have to work hard just to get back to where you were.

9. Visible Unwavering Support

It's easy to get sidetracked with all the things vying for your attention. It's easy to forget about the long-term when there are some many urgent things to deal with. It's easy to lose enthusiasm for something that's taking so long to complete. However, if *your* support is not *always* visible, other people will turn away, particularly those who don't want it to succeed. What's worse, they'll convince the uncommitted that you're not interested, and then, the project will effectively be dead. If you don't show your unwavering support, the project is likely to fail.

10. Create Short-Term Wins

People get discouraged when they're putting in the work but don't see results. That's why it's important to design the project so there are some *early wins.* That way, the managers and team leaders can point to those achievements as a taste of the improvements to come.

Usually, it's a good idea to identify areas that people have been complaining about and solve those problems early in the project. That may also free up time so they can contribute more. However, that can sometimes be counter-productive, particularly if the long-term goal is more difficult to visualize. In that case, when you solve the immediate problems, people feel there's no need to do any more. They're quite happy with the (partial) improvements. If you feel that could be the case, it might be better to leave the most acute problem until the very end.

11. Consolidate Improvements

Once a change has been proven it must be fully documented and then implemented across the organization. The team should then monitor results to make sure it's fully embedded in the process before moving on. Where possible, each change should provide a foundation for other improvements so that the process continues to evolve.

One temptation you'll face is to *declare victory too early.* There's a lot of work needed to make sure that a change that's been proven to work, and implemented in the process, will continue to be used. The team could walk away, and the process could slowly revert back to what it was before the project. Even at the end of the project, you should make sure there are periodic checks to prevent regression.

The Whirlpool

As we've seen, there are many dangers facing you as you introduce change. The most acute is the constant whirlpool that you face every day. All the things you must do, from the routine to the trivial to the urgent, will divert your attention from the long-term. Of course, long-term work is never urgent. However, many of the problems you face today are due to the lack of a long-term solution. Eventually, if you don't address long-term issues, you'll be faced with a crisis – and by then, it may be too late to solve it.

The Best Time to Change

The best place to introduce change is on a *greenfield site*. There are no bad habits. Everyone is new and willing to follow any direction you give them. And yet many CEOs just throw away that wonderful opportunity. They want to get a new site up and running as quickly as possible, so they copy work practices from existing sites. They get experienced (and biased) people from those sites to train the new recruits. Before long the opportunity is gone, and the new site has all the problems of existing ones. A greenfield site is the ideal opportunity to take the best practices from existing operations and combine them with the innovative changes you want to introduce.

FURTHER READING

On Change Management, *HBR'S 10 Must Reads*, Harvard Business Review Press.

17 RISK

What is *risk*? It's the possibility that something (usually negative) will happen in the future. Risk is always related to *probability*. If you buy a lottery ticket or invest in a new venture, you *risk* losing that money and there's a certain probability it's going to happen. Of course, there's risk in *everything* you do. Even if you stay in bed with the blankets over your head, there's a risk that the building will collapse and crush you. Hopefully, that risk is very low, but it's not zero – buildings *do* collapse – and if you live in an earthquake zone, the odds go up considerably. When you drive to work, there's a risk that someone will crash into you. Again, that risk is low, but it's higher than being crushed by a building.

Probability

We live in a *probabilistic* world. There's a certain probability that anything that's possible will happen. Some things have a high probability. It's highly likely that it'll rain within the next month (unless you live in a desert) – perhaps there's a 90% chance of that happening. It's highly unlikely that a meteorite will destroy your house. That probability is estimated to be around 1 in 2,200,000,000,000!

Sometimes we can insure ourselves against risk. Let's say you invest in a policy that pays out if your competitors bring a better product to market in the next year. Would that remove *all* risk? No. Even if you're covered for that risk, a *new* competitor could come along, or a *substitute product* could be introduced. It's even possible that your insurers could go bankrupt just before you make your claim. Risk is a normal part of doing business.

Risk and Insanity

You may have come across the following maxim, attributed to Albert Einstein: "Insanity is doing the same thing over and over again and expecting something different to happen." Do you think it's true?

Write down your answer before reading on (Q17-1).

It sounds reasonable, but it's not true. The reason is that the environment doesn't remain constant; it's always changing, even if you're not. Thousands of companies have failed because they continued to do the same thing, day after day, while changes in the outside world were making their business model obsolete. A better motto for CEOs is: "Insanity is doing the same thing over and over again and expecting the same thing to keep happening."

Threat Assessment

In fact, Einstein's version isn't valid even if you exclude the external environment. Your employees are providing a service or making a product, but how consistently do they do it? Are all the procedures being followed? Are people taking shortcuts? How different is your product or service compared to a week, a month or a year ago? And your buildings and equipment are getting older – is that causing problems? Your organization is at risk from many potential threats from internal and external sources. One way to tackle the danger is to use a *threat assessment matrix* to identify potential problems, rate them and then address the most critical.

Assessment

Start by setting up a team in each area. It should have four to six members and its first job is to identify all the categories in the area. Categories could include things like buildings, equipment, software, utilities, personnel, etc. Next, break down each category into the groups or items affected and then the identify the potential threats. There may be many items for each category and many threats for each item.

The following example shows this:

CATEGORY	GROUPS/ITEMS AFFECTED	POTENTIAL THREAT	IMPAC FOOTPRIN
Personnel	Programmers	High Attrition	Local
Personnel	Programmers	Flu / Colds	Local
Personnel	Programmers	Floods / frost	Natio
Personnel	Programmers	Security Breach	Natio
Software	Development	Virus / Ransomware	Loc
Software	Development	Virus / Ransomware	N
Software	Development	Virus / Ransomware	N
Utilities	Electricity	Supply Outage	Regi
Utilities	Electricity	Supply Outage	Na
Utilities	Electricity	National Strike	N

Next, look at the impact footprint of the threat. For example, is it limited to a single department, a single location or does it have a wider impact? Then identify if it might be a single point of failure (SPOF). This is where a *single* issue can cause a widespread system failure. Next, decide if the failure could have a short, medium or long-term effect.

IMPACT FOOTPRINT	SPOF	DURATION	PROB
Local	No	Long	0.5
Local	No	Medium	0.3
National	No	Medium	0.4
National	Yes	Medium	0.4
Local	No	Short	0.6
National	Yes	Medium	0.5
National	Yes	Long	0.5
Regional	No	Short	0.8
National	Yes	Short	0.6
National	Yes	Long	0.3

Finally, in this section, estimate the probability of the threat occurring. The probability is rated from .1 (very unlikely) to .9 (almost certain).

ROB	IMPACT	SEV
0.5	Loss of knowledge/experience in key areas	7
0.3	Missing Staff / Increased backlog	5
0.4	Missing Staff / Increased backlog	6
0.4	Security of finished product	9
.6	Cannot access local development system	4
0.5	Cannot access national dev. system	6
0.5	All records/documents lost	4
0.8	Loss of development services	8
0.6	Loss of development services	8
0.3	Loss of development services	9

The next step is to identify the impact and the severity of each failure. You'll normally select the worst-case scenario for each threat, but it must be reasonable. For example, it's reasonable to assume that a software virus could steal information or lock all your files. It's not reasonable to assume it'll blow up the building. Severity is rated from 1 (minor issue) to 9 (critical):

Now you'll want to see what can be done to address these. *Mitigation* means finding some way to reduce the impact of the threat (assuming there's no way of avoiding it). *Avoidance* means identifying ways to prevent it happening in the first place. *Alternative* means coming up with other ways to avoid the problem.

SEV	MITIGATION (Reduce Impact)	AVOIDANCE (Prevent Occurrence)	ALTERNATIVE Alternative Solutions	RA
7	Cross training	Hiring Conditions	Additional support groups	
5	Development transferred to India	Inoculations provided		
6	Analysts work from home	Set up home networks	Shadow team in China	
9	Limit access to sub-module level	All code has ID tracer		
4	Triple back up-system / virgin equipment	Anti-virus / Good practice		
6	Triple back up-system / virgin equipment	Anti-virus / Good practice		
4	Hard - copy (Locked Disk) backup system	Anti-virus / Good practice		
8	Work from home	UPS / Generator	Local company facilities	
8	UPS	UPS / Generator	Local company facilities	
9	Generator	Generator	Local company facilities	

The probability and severity columns are multiplied together to get an overall rating. This figure can be used to prioritise your response. If a threat could have serious consequences and is likely to happen, then the rating figure will be high and you'll need to tackle it. However, you can't just close your eyes to the other threats. Some of them may also need to be addressed quickly, despite a low overall rating.

ALTERNATIVE Alternative Solutions	RATING	EFFORT
Additional support groups	3.5	7
	1.5	7
Shadow team in China	2.4	3
	3.6	7
	2.4	3
	3	3
	2	6
	0	
Local company facilities	6.4	3
Local company facilities	4.8	8
Local company facilities	2.7	8

Some threats have a high impact but a low rating because they rarely occur. However, looking at your own history isn't always the best indicator of what's likely to happen. For example, if you haven't been affected by *ransomware,* you're likely to rate it as a low possibility. However, you should supplement your personal experience with the external view. When you do that, you'll find that 42% of all firms have been affected. That's the *base rate* – it should be the factor you use to estimate the probability, not your own zero experience.

There's one final column to fill out – the effort column. This is an estimate of the difficulty of implementing a solution. Some threats may get a relatively low rating, but the solution is so easy that it makes sense to eliminate them quickly. This column also gives an indication of the effort need for important items.

Recording

Make sure that the team records the reasons for their decisions as they complete the matrix. If they select an occurrence of 6 for something that has never happened, you'll need to understand how they arrived at that figure. You'll also need to be sure there was a robust discussion about the issue and people didn't just blindly accept what others were saying.

Implementation

The threat assessment matrix is a great start – it forces people to look at areas where the organization is vulnerable. However, it's important to *work* on the findings. There should be an ongoing programme to implement the solutions and it should be reviewed on a regular basis – at least twice a year. The matrix should also be reviewed regularly.

Risk From Products and Services

The threat assessment matrix is used to examine your organization and flag serious risks, but what about the risks associated with your products and services? You can use a technique called Failure Mode and Effect Analysis (FMEA) to analyse those risks. It's used to examine different ways your offerings can fail and provides a method to prioritise remedial action.

Incidents

Suppose you supply food products to consumers, and there have been three reports of customers finding metal objects in their food last year. None of the problems were serious, the items were identified before the food was consumed and there was no adverse publicity. Your review team are now discussing the incidents. One half of the team think that the issue should be rated at 9 for severity. The other half point out that the data has shown that those problems are not serious and suggest a rating of 3. What do you think?

Write down your answer before moving on (Q17-2).

Near Misses

When *near misses* occur, it means *something significant has gone wrong* but you were lucky that the problem was contained or that the consequences were not as severe as they might have been. However, imagine the results if someone had died from eating one of those objects. The fact that something like that didn't happen was *outside your control* – it was a lucky break that you can't depend on next time. When a near miss occurs, it means that your systems are not good enough – you should immediately find the root cause of the problem and fix it.

Common examples of near misses include:

- Mistakes found before they can do any damage
- Systems that need constant adjustment
- Customer complaints
- Re-occurring problems

Never get complacent about near misses because, eventually, a serious result *will* occur. It's just a matter of time and probability – and you can't say you weren't warned.

Base Rates

We've already talked about taking the outside view. This means looking at the statistics for a situation, based on data from other organizations. You can also look for base rate data from internal sources if you're sure the data is reliable. For example, suppose the engineering manager has outlined a project to introduce a new product line. She's estimated that it will take fifteen weeks to get the line into production. Is that realistic? She's adamant that she can do it. However, you don't want to present to the board and then find she's wrong. *The answer is to look at similar projects that you've already completed.* When you consult your records, you find twelve similar projects. The average completion time was forty-eight weeks. The shortest time was thirty-nine weeks. However, the average *estimated* time was nineteen weeks, with the shortest estimate being just ten weeks. Why such a huge discrepancy?

This is the *"what you see is everything there is"* bias again! The engineering manager can imagine the main work items that need to be completed, but not all the small, hidden tasks, and she's ignoring the hundreds of things that could go wrong. Without realising it, she's assuming that everything will go perfectly, that all resources will be available when she needs them, that nobody will be out sick, nobody will leave, and everything will work first time. In other words, she's made some very unrealistic assumptions. It's lucky you had accurate records! Without that knowledge, you might have promised your board something that couldn't be delivered. This is an example of using information to reduce risk.

Information and Risk

When you make a decision, you're often basing it on your memory of historical events. You're assuming that the same kind of things will happen again. That's a risky assumption, and it's even riskier because you've an imperfect memory of what happened. We've seen how your brain "massages" your memories to make them more acceptable. Those memories often omit pertinent information, like the actual time to complete the project or the fact that four of the twelve projects ended without achieving their objectives!

If you don't keep accurate records of previous decisions and their outcomes, you're increasing the risk of failure. All future decisions will be based on imperfect knowledge and you'll never learn from your mistakes.

Black Swans

Some events are so unlikely that they'll never appear in the risk assessment matrix. They've been called "black swans" by Nassim Nicholas Taleb. While these *could* have a radical effect on the organization, they're impossible to predict – so what can be done? One answer is to avoid making the organization too "brittle". If you increase efficiency to the point where it's incapable of dealing with unexpected events, then the organization will flounder when one arises. For example, if a telephone company reduces its service crew to the bare minimum required for *normal* conditions, and then an exceptional storm hits, customers could be without service for months.

Risk Appetite

Let's say you've been offered an investment opportunity. You've carried out due diligence and confirmed that it's legitimate. The essential details are as follows: you have a 60% chance of doubling your investment within 10 months and a 40% chance of losing it all. The investment is €30 million. However, that's your entire investment fund. Think carefully. Would you invest in the proposition?

Write down your answer before continuing (Q17-3).

Now, let's pretend you said "no" to that, but you've been offered an alternative. The details are the same as last time but now you're committing to investing €3 million once a month for 10 months. Each monthly investment is independent of the others. Would you invest? Think carefully.

Write down your answer before continuing (Q17-4).

First, let's see what the expected value of the proposition is. To find it, just multiply the return by the probability of it happening and subtract the investment. In the first case: (60 x 0.6) – 30 = 6 million. The expected value is positive and significant, so the proposition makes sense. However, you're betting everything on a single investment. *If you said "yes", you've got a high appetite for risk.*

What about the second case? This time you're investing €3 million each month for 10 months. Over the 10 months, the expected return is the same as the first example but this time the risk is much lower. *If you didn't say "yes" to this proposition then you have a low-risk appetite.* There's a fundamental difference between these two cases. In the first case, the expected value doesn't really tell you what's going to happen. You'll either end up with €60 million or nothing. While that's an excellent return, and the odds are in your favour, there's a significant downside risk.

In the second case, you're highly unlikely to end up with zero (probability: 0.01%) or €60 million (probability: 0.6%). Instead, you could expect to come out with something close to €36 million. Of course, that's not guaranteed but, as we've seen, nothing ever is.

Your Appetite

What do these examples say about you? If you have a high appetite for risk then it's likely that you'll take unnecessary chances. You'll bet on things that you shouldn't, or you'll commit everything on a single gamble. You're likely to go out with a "bang".

If you have a low appetite for risk, you'll avoid even small amounts of risk. You'll stick to the status quo and "guaranteed" options (with negligible paybacks). You'll pass up opportunities that could significantly benefit your organization because you're afraid of the potential negative consequences. You're more likely to go out with a "whimper" as your capital is slowly eroded.

While there's no perfect answer to the question "what's the right level of risk appetite?" we can use the 4-1-1-4 tool to indicate the sweet spot and show that, for CEOs, too little risk can be as bad as too much.

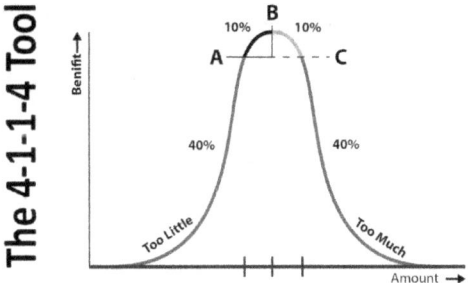

Risk and Consequences

The last example demonstrated that the absolute amount you risk is not always the issue, it's the relationship between the amount you have and the amount you gamble. Let's look at an extreme case: you've lost all your money, you've also lost your job, your spouse has left you, and the bank is coming to repossess your house. You have zero chance of ever getting another job. However, you've been given a chance to win €50 million by an eccentric millionaire. All you have to do is play one game of Russian roulette and survive. The gun has six chambers, and only one will be loaded, so you have one chance in six of being shot. The odds are in your favour – would you risk it?

The situation is dire, so you might be tempted – but one chance in six of being killed? You'll probably say "no". Now you get more information about the gun. The chamber is finely balanced and the weight of the bullet influences where it stops. The odds of it stopping in the wrong place are actually 400 to 1 against. So the odds are massively in your favour. Would you take the chance?

Write down your answer before continuing (Q17-5).

The Stakes

The trouble is that the stakes are still very high – you only have one life. Nevertheless, if you're desperate enough, you might do it. After all, what else *can* you do? So, let's pretend for a moment you say "yes". You *are* desperate enough to risk the only thing you have left. Now let's make one change to the scenario. You just remembered that you haven't lost *all* your money. You've €250K stashed away in an offshore account (I'm not judging). Are you more or less likely to take the risk now?

This thought experiment shows that your reaction to risk is strongly influenced by your current situation. In the last scenario, there was no change to the gain (€50 million) or the odds (400 to 1 against) or the wager (your life) but your decision probably changed. The same principle also applies in less extreme circumstances. For example, a CEO who's just lost €550 million and has only €30 million left is much more likely to risk that €30 million on one grand gamble with long odds and a big payback than someone who's built their company up and just accumulated €30 million. That's illogical, but it demonstrates how System 1 influences our risk assessments.

Risk Intelligence

Are you honest with yourself about what you know and don't know? Do you often fool yourself into thinking you know more than you do? If so, it's likely you have poor *risk intelligence*. That means you're more likely to avoid good investment opportunities and invest in risky ones, and you'll do so not because of your risk appetite, but *because you can't distinguish between risky and non-risky situations.*

Dylan Evans, a British academic and author, created a test to check an individual's risk intelligence. The test is used to estimate a person's Risk Intelligence Quotient (RQ). This is a measure of their ability to estimate probabilities accurately. If you score highly on that test, you're more likely to make better predictions than those with low RQ.

Ambiguity Intolerance

One trait that could reduce your RQ is *ambiguity intolerance*. This is where you come across a situation you haven't encountered before and doesn't fit into your existing knowledge maps. If you're intolerant of ambiguity, you'll react in black or white terms. It's either terrible or magnificent. You'll also react emotionally, with unease, anger, discomfort or anxiety. Those responses mean you won't be able to assess it impartially and you're more likely to make an error of judgement.

Improving RQ

There are several steps you can take to improve your predictive ability:

- Establish your current RQ
- Keep track of information in a methodical way
- Seek diverse sources of information
- Make forecasts
- Specify a level of uncertainty using percentages
- Get feedback on your forecasts

Establish Your Current RQ

Are the following statements true or false? Try to recall every scrap of information you know about each one (without looking it up!) and then decide:

- Albert Einstein failed maths at school.
- Algeria is the largest country in Africa.
- The tallest man in the world is over 8ft 2in tall.
- Alcohol sinks in water.

Write down your answer before continuing (Q17-6).

Now, estimate how *sure* you are about your answers. If you're *certain* an answer is true, give it 100%. If you're *certain* it's false, give it 0%. If you're totally unsure, give it 50% – but try to avoid that rating. For example, if you believe the second answer is false, but you're not certain, you might give it 20%.

Record your results before reading on (Q17-7).

If you do lots of these evaluations, you'll build up a picture of how you rate situations where the answer isn't clear cut. If you tend to stick to 0%, 50% and 100% you're *intolerant of ambiguity*. If that's the case, you should practise becoming more tolerant. Alternatively, you could get other people in the organization, who're naturally tolerant, to do this job, but you'll have to depend on their judgement.

The next step is to determine if you're *over-confident, under-confident,* or have the *right level* of confidence. The way to do that is by looking at all your predictions for a specific confidence rating and determine how many are right. For example, if you gave 30 statements a rating of 20%, you'd expect that six of them would be correct. Remember, you believed that *all* these statements were incorrect but you weren't certain, so you said there was a 20% chance that they were right. Suppose it turns out that 10 of them were correct. That means you were *more* confident than you should have been – you should have rated them at 33%.

The following graph shows what over-confidence, under-confidence and the correct level of confidence look like (the diagonal line indicates a perfect level of confidence):

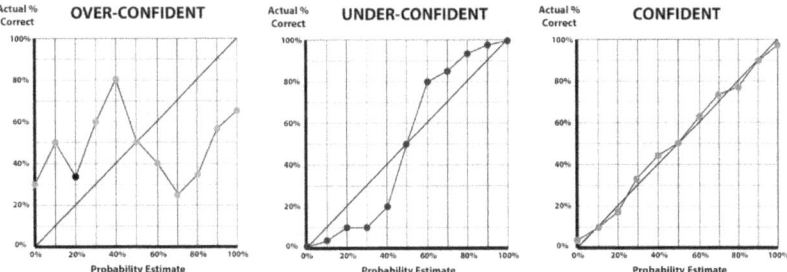

Keep Track of Information

If you don't have a structured storage system, then relevant information may be lost. You might completely forget it, have a vague recollection that you came across it or misremember it. If you *use* information like that, you have no idea how reliable it is. For each of the following statements, which are correct, and can you remember where you discovered that information?

- The bible says that Eve ate an apple and was cast out of Eden.
- People use just 10 percent of their brain.
- Bats are blind.
- A spacecraft heats up on re-entry because of friction.

Record your results before reading on (Q17-8).

All these statements are incorrect. Did you believe any of them? If so, you've accepted misleading information. The same System 1 mechanisms that allowed you to believe *those* statements are also at work for business-related information – including the most critical. You need a *system* for keeping track of information. By the way, did you say those examples aren't important – just trivia? By now you know that's your brain trying to protect your ego.

Seek Diverse Sources of Information

The same nugget of information – true or false – is often repeated by numerous sources. But those may not be diverse sources, just several sources re-broadcasting from a single source. It takes determination to find alternative sources offering independent information. Of course, the validity of the information must still be *checked* before allowing it to influence your risk management process.

Make Forecasts

It's possible to improve your risk management skills but it's like any skill – you have to practise to get better. So, start making forecasts. Write them down and make them *specific*, *measurable* and *time-based*. For example: "The Chancellor of the Exchequer is going to raise VAT on alcohol by at least 5% before the end of November." It's important to stress – this *can't* be just a guess. It must be based on your analysis of all the information you've been able to gather about the area.

Some people have a problem writing down anything that'll *clearly* prove them right or wrong. They prefer some "wiggle room". That's why you'll often see pundits add phrases like "assuming things continue as they are…" Of course, that's ridiculous, because *things never continue as they are*.

If *you're* reluctant to write down your prediction in clear and simple language, then your System 1 is trying to prevent you noticing your mistakes. You'll need to overcome that reluctance if you want to improve.

Specify a Level of Uncertainty

It's not enough to make a prediction. You also need to specify how sure you are that it will happen. Use the system discussed earlier where 50% means unsure and 100% is absolutely certain. For example: "The Chancellor is going to raise VAT on alcohol by at least 5% before the end of November – with a 90% probability."

Get Feedback

Who makes the most accurate predictions, doctors or weather forecasters? Let's consider the evidence:

Doctors

- They deal with a range of domains (accidents, diseases, etc.)
- They don't express their diagnosis with a probability value
- They often don't get feedback

Weather forecasters

- They operate in a single domain (is it going to rain)?
- They usually express their diagnosis with a probability value
- They get rapid feedback

As you might expect, weather forecasters are significantly more accurate than doctors. In studies, when weather forecasters estimated there was a 90% chance of rain the following day, it rained almost exactly 90% of the time. When doctors estimated there was a 90% chance of their patients having pneumonia, only 15% of them had it! Doctors were massively overconfident about their diagnosis. You might think it's unfair to compare weather forecasters and doctors because their jobs are so different. That's a fair point – but do you think it's reasonable for *any* profession to be that overconfident? Particularly when the consequences are so serious?

And this brings us back to *your* forecasts. Are you more like the weather forecasters or the doctors? If you don't have a feedback system in place, you might be *worse* than the doctors – there's no way to tell. If you want to improve your risk management, an effective feedback system is essential.

Working With Risk

So how do you decide when there's risk involved? One approach is to separate the process into two parts. First, decide what's an *acceptable* level of risk for you and your organization. Second, estimate the risks involved in the various options you and your team are considering.

Acceptable Risk

Get your management team involved when you calculate the acceptable level of risk for your organisation. There are four questions you should consider:

- What is the current situation?
- What percentage of your capital are you considering investing?
- What is the payback?
- What level of risk are you happy with?

The Situation

In theory, your current situation shouldn't affect your investment decision but – as we've seen – it does. That's why you should at least make the decision explicit. Here are three possible situations you should consider (you may want to add more):

1. You're doing OK and your available capital is holding.
2. You're losing money and your available capital is decreasing.
3. You're doing very well, and your available capital is growing.

How will these situations affect your risk? Let's assume situation 1 is your reference point – the norm. If you're facing situation 2, it's likely you'll be *more* inclined to take greater risks for a big payback. You and your team need to agree on a figure that will apply in those situations – and that becomes the rule. It's important to reiterate that *this is not logical*, but having the rule in place may prevent an even greater risk being taken when facing the situation.

What about situation 3? When things are going well, there's more money available and logically, you should be willing to accept riskier investments. However, the reverse is often the case. If so, it might be prudent to specify that you will – at least – accept investments that are somewhat less risky than normal.

The Percentage of Capital

If you're risking 1% of your available capital, and it doesn't work out, it's a disappointment but not a disaster. You can still make another 99 similar investments. However, if you risk 100% of your capital, you only get one chance and that's more like a gamble than an investment. So, the greater the percentage you risk, the better the odds need to be. Many small independent investments will return a value close to the expected value, assuming they all have roughly the same probability of success (ideally in non-overlapping areas).

The Payback

Let's say the yearly interest rate on government bonds is 5%. That's as close to risk-free as possible. There's little point in investing in a project that has *any* chance of failure if the return is going to be 5% or less (ignoring the cost of having money locked up). On the other hand, let's say there's a chance of getting a 500% annual return on your investment. Then you might be willing to take a significant risk. So the greater the return, the greater the risk you'd be willing to take.

The Risk

The risk you're willing to take in any project should factor in the items discussed above. The following is an example of how it might look:

NORMAL RISK

ACCEPTABLE RISK (risk of failure)

Percentage of Capital

Annual Return	1%	5%	10%	20%	40%	60%	85%	90%	95%	100%
5%	-	-	-	-	-	-	-	-	-	-
10%	20%	15%	10%	5%	3%	2%	-	-	-	-
20%	25%	20%	15%	10%	5%	4%	2%	1%	-	-
50%	30%	25%	20%	15%	10%	6%	4%	2%	1%	0.50%
75%	35%	30%	25%	20%	15%	8%	6%	4%	2%	1%
100%	40%	35%	30%	25%	20%	10%	8%	6%	3%	1.50%
250%	50%	40%	35%	30%	25%	12%	10%	8%	4%	2%
500%	60%	50%	40%	35%	30%	15%	12%	10%	5%	2.50%

You can see from the "normal risk" matrix that you'd be willing to invest in a project that gives a 10% annual return if it has an 80% chance of succeeding (20% chance of failing), providing that it requires 1% or less of your available capital. However, if a project required 100% of your available capital, you wouldn't invest unless it had at least a 50% annual return and it would need at least a 99.5% chance of success.

The figures in the "high risk" matrix might surprise you:

HIGH RISK

ACCEPTABLE RISK (risk of failure)

Percentage of Capital

Annual Return	1%	5%	10%	20%	40%	60%	85%	90%	95%	100%
5%	-	-	-	-	-	-	-	-	-	-
10%	20%	15%	10%	5%	3%	2%	-	-	-	-
20%	25%	20%	15%	10%	5%	4%	2%	1%	-	-
50%	30%	25%	20%	15%	10%	6%	4%	2%	1%	0.50%
75%	35%	30%	25%	20%	15%	8%	6%	4%	2%	1%
100%	40%	35%	30%	25%	20%	20%	18%	16%	13%	11%
250%	50%	40%	35%	30%	25%	22%	20%	18%	14%	12%
500%	60%	50%	40%	35%	30%	25%	22%	20%	15%	13%

There's no change to the figures on the left of the matrix. It's only at 60% of capital and higher that the increases kick in. The reason is that CEOs in a difficult situation are usually not interested in small gains. They want a large absolute amount of money to solve their crisis. The low-risk matrix is as follows:

LOW RISK										
	ACCEPTABLE RISK (risk of failure)									
	Percentage of Capital									
Annual Return	1%	5%	10%	20%	40%	60%	85%	90%	95%	100%
5%	-	-	-	-	-	-	-	-	-	-
10%	15%	10%	5%	-	-	-	-	-	-	-
20%	20%	15%	10%	5%	-	-	-	-	-	-
50%	25%	20%	15%	10%	5%	1%	0.50%	-	-	-
75%	30%	25%	20%	15%	10%	3%	2%	2%	1%	-
100%	35%	30%	25%	20%	15%	5%	4%	3%	2%	0.50%
250%	45%	35%	30%	25%	20%	7%	6%	4%	3%	1%
500%	55%	45%	35%	30%	25%	10%	8%	6%	4%	1.50%

Definitions

There are several things we haven't discussed: first, what *is* available capital? You might define it as your current assets minus your current liabilities. Alternatively, you might include all the available money – including borrowings – that you can lay your hands on. Write down the definition you're using because it'll make a difference to the risk figures.

Second, we're assuming that you'll lose *all* of your investment if the project fails. That isn't always true. If it can be guaranteed that some portion of your investment will be recovered, you can subtract that from your initial investment when calculating the percentage of available capital. We've also ignored the timescale of the project. One project might give an annual return of 5% and be completed within a year. Another might give the same annual return, but it might not be completed for five years. You may need a separate matrix for each timescale you're likely to face.

Risk Rules

You and your team should create an acceptable risk matrix inde-pendently of any actual projects in the pipeline. The figures should not be influenced by current requirements or desires. Once completed, the matrix should become the investment rulebook for your organization.

You should *never* break the rules – particularly to take advantage of a "once-in-a-lifetime-opportunity". The matrix is there to protect you from impulsive (System 1) mistakes, but it'll only work if you stick to it when your gut is telling you to do otherwise.

Calculating Project Risk

The second part of the risk equation is calculating the risk associated with a project or investment opportunity. The best way is to break it down into elements and calculate the probability of each of those occurring. Use all available information, including internal and external base rates, to estimate probabilities. Then calculate the overall probability of success by multiplying the individual probabilities together.

For example, suppose you want to create a new product. You don't have a lot of information, but you want to estimate the probability of success before going too far. Let's look at the possible risks in one area:

- Packaging will be too costly – probability: 0.2%
- Distribution will be too costly – probability: 0.25%
- Large stores won't stock the product – probability: 0.3%
- Advertising will be too costly – probability: 0.5%

Assuming these are independent risks then the probability of success has already dropped. Let's assume that when you include *all* project risks, the probability of success drops to 94.5%. Should you go ahead – what do you think?

Making the Decision

You need two additional pieces of information: the amount of capital required (10% of available capital) and the projected annual return (20%). This is no longer a subjective decision. Look at the risk matrix (normal risks) and you'll see that the maximum risk for that combination is 15%. The risk figure for this project (5.5%) falls within the parameters, so you can continue to work on the project.

Next Steps

As you continue to work on a project, you might find you were too optimistic (or pessimistic) so you should update the probabilities as you go along. However, if the risk increases above the relevant figure in the risk matrix, you must stop the project – and that will be difficult. You'll have put a significant amount of work into the project by this stage and you're understandably reluctant to let go. This is the *sunk cost fallacy* at work and it demonstrates, once more, that System 1 is in charge. Step back and accept that any money, time and effort you've already invested are gone. The decision you face now is whether you'd invest if this was the first time you had seen the project. If the risk figure is too high, then you don't have a choice. You can only *manage* risk if you're willing to walk away.

FURTHER READING

Dylan Evens, *Risk Intelligence: How to Live With Uncertainty*, Atlantic Books

18 STRATEGY

Here's a question to test your knowledge of strategy: If you have a great strategy, are you more or less likely to disappoint some of your customers?

Write down the answer before continuing (Q18-1).

The Five Forces

We've mentioned Michael Porter and his five forces model already. The model identifies the following threats:

1. The threat of existing competitors
2. The threat of new entrants
3. The threat of substitute goods or services
4. The bargaining power of suppliers
5. The bargaining power of customers

Let's see how these forces work to limit your options and your profitability.

(1) Existing Competitors

Existing competitors constrain your actions. If you drop your prices to get more customers, they're likely to react and drop theirs, so you end up with the same number of customers but lower prices. If you increase your prices, there's no guarantee they'll do the same, and you could end up losing customers.

(2) New Entrants

If there are substantial profits in your industry, it's likely that outside companies will be attracted to it. Some might come from other industries and bring more efficient production or delivery methods. Some could harness new technologies that will negate many of the advantages you've built up.

(3) Substitute Goods or Services

Customers may decide to abandon your products or services because there's something new that satisfies their need. Typewriter manufacturers in the early 1980s didn't realise that their main competitors would be computer makers. Hotel owners in 2008 didn't expect competition from companies like Airbnb who don't own any property.

(4) The Bargaining Power of Suppliers

Each of your suppliers wants to make as much money as they can from you. If some are large multinationals with many customers and few competitors, they'll have significant power to dictate terms and prices. You may have no choice but to accept price hikes, even though they'll significantly impact your profitability.

(5) The Bargaining Power of Customers

Each of your customers wants to get the best deal possible. Larger companies, with many options, can dictate what they're willing to pay and you may have to accept what they're offering. Even small customers can become powerful by banding together and using their combined purchasing power to make a deal.

Challenges

Your strategy needs to take these forces into account, but that isn't always easy. How can you prevent new entrants before they appear? How can you prevent substitute goods or services from attracting your customers? And how do you increase your bargaining power? You might be lucky to have found a position where those forces are relatively weak. This'll help in the short term, but you can't depend on it continuing indefinitely. So what can you do? It's widely accepted that you need to develop and implement a powerful strategy if you want your organization to succeed and thrive in the long term. It's interesting to see how many organizations fail to do so – and suffer the consequences.

Example

The CEO stands up in front of the workforce. Then he moves to the microphone on the makeshift stage and starts to speak:

"I'm speaking to you today to unveil our new strategy. We're going to be a billion-dollar company and Number 2 in the industry within two years. We'll achieve that by driving excellence through all our operations using our new Total Quality Management programme. We'll take a price leadership position across our five primary product lines and increase our penetration in the European and Asian markets by adding hundreds of local salespeople."

Would you say that's a good strategy? OK, you've probably spotted that it's not a strategy at all. Let's break it down:

- The intention to become a billion-dollar company and Number 2 is a *goal*, not a strategy.
- Total Quality Management is *not* a strategy.
- Price leadership is unlikely since they're only aiming to be Number 2 and it's *not* a strategy.
- Increase penetration – that's an operational issue, not a strategy.

So, there's not a single hint of a strategy in the CEO's strategy statement. This is a common issue.

Definitions

One of the problems is that the word "strategy" is frequently used in situations where it's not appropriate. Let's have a look at a definition:

"Strategy is a high-level plan for competing in an uncertain environment to achieve significant long-term goals."

So strategy is *not* a goal or objective, it's not a vision statement, a mission statement, a values statement or a tactic. Let's define those terms to eliminate any confusion:

Vision Statement

A vision statement describes an organization as it will look when it has achieved all its strategic goals. The vision looks ahead by 5 or 10 years and provides a signpost for the future. It should motivate and inspire employees.

Mission Statement

A mission statement identifies the organization's purpose. *Why* does it exist? It may also include a description of its priorities. It should support the vision statement and may change to reflect the current areas of focus required to achieve the vision.

Values Statement

A values statement outlines what the organization *believes in* and how its employees should *behave*. It provides a moral compass for management when they're making difficult decisions. It documents the core principles of the organization.

Goals and Objectives

A goal is something to be achieved. Frequently, the terms "goal" and "objective" are used interchangeably. However, "goal" is also used, in a more restricted sense, to mean aspirational achievements like profits, employee satisfaction, etc. The word "objective" is then used to indicate actionable targets. When that distinction is made, *"goals" act as a constraint on strategy or tactics*. For example, if an organization has a goal of increasing the satisfaction levels of all customers, it may not be possible to rationalize the product range because of its impact on some customers.

Strategic Goals and Objectives

Strategic goals and objectives are achievements that will make a significant difference to the competitiveness of the organization. It may be necessary to achieve many short-term objectives to achieve a single strategic objective.

Tactics

Tactics are short-term, low-level plans for achieving objectives that (usually) support the strategic plan.

The Reality

The reality is that most vision, mission and values statements are worthless. At best, they state the glaringly obvious. At worst, they use jargon and convoluted logic to hide a lack of critical thinking or real strategy in the organization. They may seem harmless enough but the CEOs and managers who've crafted these statements believe they've achieved something, and that sense of achievement absolves them from doing any work to create a real strategy.

Mission statements are useful only if they make a difference to the behaviour of the people in the organization. They *can* serve a useful purpose when they guide strategy, but sometimes that can be counter-productive. Can you spot the difference between a good and bad mission statement? Have a look at this one:

"We provide low-cost air travel between capital cities in Western Europe."

Is that a good mission statement? Think about it for a minute.

Write down your answer (Q18-2).

When you read a mission statement, the first question is: it clear and jargon free? This one is – there's no confusion about what it means. The next question is: does it exclude anything? This one does. It excludes flying to any location outside Western Europe. It excludes servicing any city that isn't a capital. It also excludes providing a premium service.

It's a *good* statement because it informs both strategy and decision making. Should the company consider opening a new set of routes to the Far East? Should they consider taking over a road haulage company? Should they be offering complimentary beverages and reading materials? No, all these options are outside their remit as defined by the mission statement. Compare it to this one:

"To create a shopping experience that delights our customers; a workplace that creates possibilities and a pleasant working environment."

However, just because the first statement is *good* doesn't mean it's *helpful*. We'll see later how it could send you in the wrong direction.

Confusion

Many CEOs confuse strategic goals, objectives and strategy. The key difference is that goals and objectives are achievements ("we're going to be Number 2") and strategy is HOW you get there. It's easy to come up with a goal; the important (and difficult) part is to come up with a great strategy. There can also be confusion between strategy and tactics. The easy way to differentiate is scale. Strategy relates to the business as a whole, while tactics are more detailed and lower level. In military terms, tactics are used to capture a town while strategy is required to take the country.

Operational Excellence

Japanese car makers like Toyota dominated the small car market in America, starting in the 70s. They succeeded because their products had higher quality and reliability and were cheaper than their American rivals. Quality experts in the US didn't believe that was possible because they used the "cost of quality" model. This illustrates the trade-off between quality and cost:

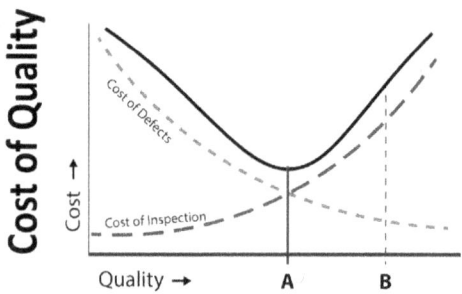

This *seems* to prove that point A is the lowest cost. If a company wanted to increase quality (for example, to B), they had to increase inspection – which, in turn, meant higher prices.

The Japanese rejected that hypothesis (influenced by W. Edwards Deming and others). They introduced a host of new quality techniques and proved that *even though the curve looked logical – it was wrong*. The new approach gave Japanese companies a huge advantage, and they exploited it to become the dominant force in the global automotive industry.

It sounds like they had a great strategy, doesn't it? Michael Porter disagrees. He argues that there's a difference between *operational excellence* and strategy. Operational excellence is necessary but it's not sufficient. The reason is that *best practice* soon becomes common knowledge and then everyone can use it. There are many operational excellence tools and techniques, including:

- TQM
- Six-sigma
- Lean Operations
- Benchmarking
- Reengineering
- Business Process Management

If an organization fails to apply these tools – and to apply them properly – it'll find it difficult to complete. However, applying them won't automatically lead to a sustainable competitive advantage.

What Strategy Isn't

So, we need to be clear about what strategy *isn't* before we define what it is and how to create and apply it. Strategy is not:

- Setting goals
- Setting objectives
- Vision, mission and values statements
- Operational excellence
- Tactics
- Action items
- Lowering prices
- Following the industry leader
- Growth
- Exhortations (e.g. positive thinking)

Of course, many of these elements may be *included* in a strategy, but, by themselves, they're not a strategy. It's like confusing a picture of a car with the car itself. All the key components are missing, and you can't *really* expect it to take you where you want to go.

What Strategy Is

So what's strategy? Here are some important elements:

- Strategy means choosing a unique position.
- It means saying to "no" to many opportunities.
- It means creating "fit" – getting activities to reinforce each other.

A Unique Position

If you're selling the same things in the same way as your competition, why should people buy your products or services? You might say because your prices are lower or your quality is higher. The problem is that your competitors are always working to lower their prices and improve their quality. If you buy a machine that reduces your costs, your competitors will hear about it and buy a similar machine. When you introduce a *total quality management programme,* your competitors can copy it and their quality will also improve. You need to find a *sustainable* advantage – something that isn't easily copied. There are some generic ways of achieving that. You could:

- Satisfy the few needs of many customers
- Satisfy the broad needs of few customers
- Serve the broad needs of many customers in a narrow market

Let's say you're thinking of opening a garage but there are many competitors – they're all attempting to satisfy the *broad needs* of *many* customers. If you were to concentrate on a single area – for example, the fitting of exhausts, then you're addressing the *few needs* of *many* customers.

Alternatively, you could concentrate only on wealthy customers. You could give them a five-star experience with pick-up and drop-off services and executive replacement cars. Now, you're serving the *broad needs* of a *few* customers.

Or you could concentrate on the *broad* needs of a *narrow* market – for example, you could concentrate on the repair and servicing of vintage cars.

Why They Might Work

So, why might these strategies work? In the first case, you gain benefits from specialising. For example, you might be able to buy exhausts in bulk, the fitters don't have to be qualified mechanics and the space requirement is much less than a traditional garage. That means costs are lower so you can charge less. Also, there's only one activity so the procedures and layout can be simplified and the time involved can be reduced. That might allow you to offer a ten-minute turnaround time.

In the second case, you can address the *exact* needs of your target group. In this example, the selected customers are not cost sensitive but they are time sensitive and demand high quality. If you can save them time and provide a personalised service, you can charge them more.

In the third case, you select a narrow market (a niche) that will allow you to develop an *expertise* and *reputation*. In this example, most modern mechanics have no experience dealing with vintage cars. There are no electronic diagnostic systems, the mechanical and electrical systems are relatively primitive and parts are hard to source. If you can create an expert group to support these machines, then you may be able to develop a captive customer base.

Why They Might Not Work

It's important to pick a strategic position that has sufficient potential. For example, if exhaust systems rarely need to be changed, then selecting this niche would be a mistake. So would selecting wealthy customers if most of them lease their cars from companies with in-house service facilities. And if most vintage car owners take pride in doing their own servicing, then you'll quickly find yourself in trouble.

When you select a position, you must also look at the wider implications. For example, when you set up a "typical" garage, you'll put up a sign and advertise in the neighbourhood papers. Most of your customers will be locals who won't have far to travel. When you set up a specialist garage, you'll need to publicise your business more carefully. You may need to advertise in specialist publications – like business magazines or vintage car newsletters – and you may even need to do a significant amount of direct marketing.

Saying "No"

You can see that strategy is about choice. It's about saying "no" to opportunities. Sometimes, that can be hard. For example, you've set up the exhaust shop and things are going well, but then you notice that many of your customers also need tyres, so you add them. After a while, you notice that many of them also need alternators, regulators, batteries and headlamps. You add all these products but now you've got to hire some qualified technical staff. You're also holding larger inventories and the turnaround time has increased to an hour. Slowly, costs and prices begin to increase, and customers are less sure about what you do. You've started to lose your strategic advantage.

Creating "Fit"

How do you get activities to reinforce each other? Let's illustrate by considering another example. It's 1981 and you own a chain of stores in the US. A salesperson is trying to sell you one of the first commercially available barcode systems. She explains the advantages: It takes less time to check out items, you don't have to put prices on everything, the system will produce bills automatically and nobody can steal from the till because the system logs every item sold. It's interesting, but will it give you a *strategic advantage*?

> **Write down your answer before reading on (Q18-3).**

You'll have realized that the answer is "no". All your competitors can buy the same system and get the same advantages, right? Well, that *should* be the case, but one of your competitors is Wal-Mart and they actually have a strategy.

Wal-Mart is a "general merchandise big box discount retailer". That means its stores are big – around 50,000 sq. ft. – and it sells a wide range of goods, from groceries to clothes to hardware at low prices. In the early 80s, it was growing, but it wasn't the largest of its kind – that was Kmart. In a head-to-head competition, there would have been no contest – Kmart had the resources to wipe out the fledgling company. So what could Wal-Mart do?

The conventional wisdom was that large stores had to locate in cities of at least 100,000 people, but Wal-Mart located many of its stores in towns of 10,000. As you might expect, they had a strong focus on keeping costs low, but they also operated their own transportation system. Their trucks delivered goods to the stores from a regional warehouse. On the way back, they picked up goods from their vendors. They also operated a central computerized system that linked every store – one of the earliest.

Let's examine those elements in detail. Why did they locate stores in towns that seemed to be ten times too small? One reason was there was no competition from big stores – they didn't have to worry about Kmart coming in and starting a price war. However, to be profitable, they had to vigorously minimize costs, and one way to do that was to reduce the amount of inventory held. That's why they used their own trucks. The inventory was kept in regional centres and transported to individual stores as needed. They also kept the workforce small so most of the administration work was carried out at headquarters.

Now let's get back to that barcode system. For most competitors, it provided some operational benefits. It reduced staff levels a little and reduced delays at the checkout, but that's about it. Wal-Mart took a strategic approach. They used the system to track stock levels of *every* item in each of their stores in real-time. It meant that trucks could be stocked with the exact needs for each store – no waste. Orders automatically went out to their vendors for just-in-time replacement of stock, so inventory levels in the regional centres were kept to a minimum. They also began to track trends across stores to predict whether an item might be on the way out. That meant they cut back on items that were coming to the end of their life. The following list summarizes how the different elements of their strategy fit together:

- They selected small towns (minimal competition).
- They offered low prices.
- They minimised local inventory – reducing costs.
- They minimized the workforce.
- They centralised their inventory.

They also:

- Controlled and optimised the distribution system.
- Used a network to centralized administration.
- Used data to optimise the supply chain.
- Used live data to make purchasing decisions.

The barcode system became a key element of their strategy, providing real-time information to support their core systems. It also provided additional opportunities. For example, they were able to get discounts by sharing information with their vendors.

The Outside-in View

It's 1985 and Wal-Mart has continued to grow. You visit some of their stores and notice they have fewer staff than you do and they're using barcode readers everywhere. You decide to open a store in a small town and use a barcode system to minimize staff. What do you think will happen?

It's not going to go well, is it? The barcode readers are *one* element of the strategy but they're not the most important. You don't have a centralised structure and adding a barcode system isn't going to change that. You don't have regional distribution centres or your own trucking fleet or a centralized administration system either. You can't even use the data generated by the barcode system in an effective way because your stores aren't on a network.

This demonstrates one of the fundamental problems with trying to copy someone else's strategy. You may *think* you know what they're doing, but you may only be looking at peripheral items and not the core strategy. If that happens, you'll go down the wrong road.

This example also demonstrates the difference between having an effective strategy and not having one. If Wal-Mart was able to move into small towns and become successful just by using operational excellence techniques or buying new equipment, it would soon face many competitors doing the same thing. However, in the real world, if a competitor wants to challenge Wal-Mart, they have to duplicate *all* its strategic elements – and that's difficult.

How to Create a Strategy

So, how do you create a strategy? Richard Rumelt, an organizational theorist and Emeritus Professor at the University of California, believes that a good strategy has a kernel – and it contains three elements:

1. A diagnosis
2. A guiding policy
3. Coherent action

(1) The Diagnosis

The first step is to diagnose what's going on and the challenges you face. One technique that can help is the SWOT analysis. You're familiar with the format:

- Strengths
- Weaknesses
- Opportunities
- Threats

Use the techniques described in the "Decision Making" chapter when you're carrying this out. Also, remember, an in-depth SWOT requires lots of information – *it can't be done over a few days*. If you try to collect and collate the information in a few sessions, you *will* overlook important areas. You need a *system* in place to build up the foundational knowledge.

The solution is to schedule regular briefing sessions for senior management on developments that could impact or energise the organization. These sessions could include information on areas like:

- Technology developments
- Political changes
- Legal changes
- Economic developments
- Social transformation
- Industry-specific updates

You can nominate specific people within the group to specialise in one or more of these areas. That'll ensure that everyone stays up-to-date and the information is fresh when you need it. Depending on your industry, it may only take one or two 60-minute sessions each week to build and maintain knowledge levels.

What are you trying to do when you carry out the SWOT analysis?

1. You're trying to simplify complexity. You're putting structure on a chaotic situation so that you and your team can understand it better.
2. You're trying to identify areas (the weaknesses, opportunities and threats) where your intervention can make a difference.
3. You're trying to find a few critical areas to focus on.
4. You're trying to use your strengths to address these areas. If that's not possible, you may need to develop new strengths.

Eyes-R-Right

Let's look at an example. You're the CEO of *Eyes-R-Right*, a chain of opticians. It was set up 10 years ago when the only competition was small owner-run opticians. Eyes-R-Right was very successful for the first eight years but profits have been slipping recently. The owner-run opticians have banded together to benefit from lower lens and frame prices and there are reports that a multinational chain is considering setting up in competition.

You've just taken over and need to create a new strategy for the business. Here's the result of your first SWOT analysis (after some trimming):

Strengths

- 68 stores
- Good store locations on high streets and in malls
- Located in towns and cities with populations over 30,000
- Good name recognition by the public (44%)
- A database of customers' details
- Two qualified optometrists in each store
- Good quality diagnostic equipment in every store
- Low customer waiting times (~10 minutes)
- Good credit rating

Weaknesses

- Return on assets (ROA) has dropped from 15% to 3.5%
- Slow turnaround time on lenses (~3 weeks)
- Generic capabilities

Opportunities

- Sufficient capital to open new stores in smaller towns

Threats

- Price pressure from local owner-run stores
- Multinational competitor(s) poised to enter the market

There were two reasons for the initial success of the business. The first was low prices compared to local opticians. The second was the national advertising campaigns that Eyes-R-Right could afford but the local owners could not. Now that the local owners have banded together, the advantage of low prices is gone. They may also start to advertise and capture even more of your customers. Consider this information carefully.

Write down your conclusions before reading on (Q18-4).

You might consider improving productivity and reducing the number of trained staff. That would reduce costs but it won't give you a sustainable advantage and it might cause other problems – like increased waiting times. You could attempt to copy Wal-Mart and open branches in smaller towns, but there are already opticians in those towns and they could quickly join with their city colleagues and lower their prices to match yours. You no longer have the key advantage that the company enjoyed 10 years ago.

At this point, many CEOs would give up on strategy and just try to improve efficiencies and pull up the ROA a few percentage points. They might even cheat by reducing the asset base. They might sell off some of the less profitable stores or work some magic with depreciation figures.

However, that's just hiding the lack of a solid strategy to deal with immediate and future threats. So, let's take a more detailed look at the important elements. Let's start by looking at your key weakness – it's labelled "generic capabilities". What does that mean?

- Your equipment is generic
- Your talent is generic
- Your knowledge and expertise is generic
- Your administration structure is generic

You use equipment that's widely available at a relatively low cost. You're using qualified optometrists who are competent at their job but have no special skills to distinguish them from others. The knowledge and expertise you use are standard for the ophthalmic industry. All your stores are linked by computer to a central administration office, but that's typical of every multi-location business. So, you have *no* strategic advantage in any of these areas.

But what about your strengths? Unfortunately, most of these are not strategic either. You have name recognition, but it's not enough to prevent your customers defecting to your competitors. If a new entrant was to advertise aggressively – particularly with lower prices – that advantage would be lost.

The locations of your stores are an advantage, but again, new entrants could buy or rent similar properties. In fact, you only have *one* strategic advantage – can you see what it is?

Before reading further, try to develop a strategy for this company. Use all your knowledge and experience to identify the best way forward. Do some additional research if you think it might help.

> **Write down your strategy and compare it to what's described in the following paragraphs (Q18-5).**

The Way Forward

There's only one thing that your competition would find it impossible to duplicate and that's your database. You have your customers names and addresses, their contact details and their history. You have ten years' worth of data that no newcomer can replicate. Of course, your existing competitors have their own records, but those are fragmented, and it's unlikely that the owners will share that information.

The database is a good starting point, but it's not enough. The problem is you're offering a generic service. You offer branded products (spectacle frames) but so does everyone else and many of your customers are probably not brand conscious. What else can you do?

This is where your *briefing sessions* are useful. Your technology expert has presented the following information:

- Dragon from Nuance Communications is a voice recognition system with "deep learning" technology. This helps it achieve high levels of accuracy. It can learn common words and phrases and adapt to background noise.
- An AI program developed by Google can examine retinal images and detect *diabetic retinopathy*, a disease that affects one-third of diabetes sufferers. It performs as well as a highly trained ophthalmologist.
- Alibaba, a Chinese e-commerce company, has launched Buy+, a virtual reality experience. Using a headset, customers can wander around a store, examine the merchandise, and add things to a cart by staring at a product.
- Artec's 3D scanners can digitize real-world objects with complex geometries. Its Eva model can scan the human face and produce a detailed 3D model in minutes.

These stories hint at a possible direction. Imagine the following scenario: A customer comes in and sits at a machine. A chatbot welcomes him, asks for some details and then asks him to hold still for a few seconds while his face is scanned. It then asks him to look into a viewer while his eyes are checked.

As this is happening, the chatbot gives him instructions, asks him questions ("Is this better or worse?") and answers any queries he might have. The machine takes several retinal scans and the AI system checks them for any anomalies.

Once the checks are complete, the system displays a virtual store with a wide range of frames. The customer can specify the materials, colours and price range. When he selects one, it appears on a 3D image of his face, so he can see exactly what it's going to look like. Finally, he chooses one and buys it using his credit card. The glasses are shipped directly to his address the next day.

That might seem like science fiction but, as we've seen, all the core technologies are already in place. Obviously, you'll have to dig deeper to make certain there aren't any stumbling blocks, but let's assume you've done your research and there are no major impediments. Now you'll have to set up a development group to integrate the technologies and design and manufacture the machines. You have the resources, but is it a good idea?

One objection might be that these machines will turn eye tests into a commodity service. People won't be willing to pay much for an eye test when it's done automatically by a machine. That's probably true, but the eye test is just a loss-leader to get people in the door. It's not the main revenue generator.

The most important question is: will the machine provide a strategic advantage? The technology will significantly reduce costs and provide a seamless experience for the customer, so it offers significant advantages. It's also unlikely that a loose group of owner-opticians will be able to develop a similar machine, so it offers a strategic advantage over *those* competitors.

However, if a multinational organization invades the market, they'll have the resources to develop a similar system. You may have an advantage for a few years while they're developing it but eventually, they'll be able to compete. You'll need another ingredient to turn it into a strategic advantage.

(2) The Guiding Policy

Now that you've analysed the issues and the challenges, it's time to create a policy that'll take your organization forward and keep it orientated in the right direction. You might work with your team to come up with something like this:

"We'll use an automated system called "StrongEye" to carry out the work of opticians. The system will also provide customers with a virtual store in which they can easily view, select and pay for spectacle frames. We'll use the information in our database to optimise the competitiveness of the StrongEye machines."

(3) Coherent Action

Now you need a list of actions that will make the strategy happen. You don't have to be too detailed but they should provide clarity about the next steps. For example, you and your team might create the following:

Strand 1

- Create a separate development company
- Hire technical experts
- Set up the development group
- Develop the StrongEye system
- Get official approvals for the system

Strand 2

- Write to everyone in the database
- Offer them free eye tests for their family
- Set up a mobile eye testing facility
- Travel to small towns and villages
- Offer a free eye test for everyone
- Advertise in large towns and cities
- Offer all customers a free eye test

The actions in Strand 1 arise directly from the guiding policy and are expected, but what about the actions in Stand 2? Can you see what they have to do with the strategy? Let's have a look at Strand 3 and see if this clarifies the situation.

Stand 3

- Test the system in selected Eyes-R-Right stores
- Deploy the system in Eyes-R-Right stores
- Negotiate an agreement with small-town pharmacies

Then:

- Deploy the system in small-town pharmacies
- Write to local customers offering free eye tests
- Deploy the system in city pharmacies
- Write to local customers offering free eye tests
- Maintain the system

The system can be tested in one of your stores to get rid of any bugs. You can also monitor whether people like using it or not. If there are problems, you can address and eliminate them at this stage. Next, you'll roll out the machine in all your stores. At the same time, you'll start to negotiate with small-town pharmacies to install the equipment in their premises. They'll get a percentage of each sale and the machine will also attract customers to their store. Once you have achieved this rollout, you'll do the same for selected city pharmacies.

Now, the strategy becomes clear. You're *not* using the machines to increase productivity in your stores – you're using them to *eliminate* your own stores (and those of your competitors).

But what about Strand 2? Well, you'll need to use the database to promote the new machines. You'll send personal letters to each person, showing them the location of the closest machines. You can differentiate those letters by including details of their last test results, explaining what each parameter means, and arranging their next session. You'll customize it by age group, making the machine seem exciting for younger people and reassuring older people that it's easy, safe and confidential. You'll promote yearly check-ups that will only take a few minutes in their own area. The database gives you a trusted pipeline directly to your customers. You won't need to depend on expensive advertising.

So, the primary purpose of Strand 2 is to capture contact details for the database. You want to get as many people on the system as possible. There's also an important secondary purpose: you want to minimise the number of people who'll be available to the multinational if it decides to compete in the market. If someone gets a free eye test from you today, it's unlikely they'll be interested in another one from a different company in three months. So, Strand 2 is *not* a once-off sales promotion, it's an investment in the future of your strategy.

The Proximate Objective

Richard Rumelt emphasises the importance of "proximate objectives". These are objectives that are feasible and achievable with a reasonable amount of effort. People in the organization can see what needs to be done and the objectives provide direction. For example, the main objective of Strand 1 is to develop the StrongEye system. We *know* the technologies exist and there are experts in each of the technologies, so this becomes the first of our proximate objectives. We've already defined the key actions required to achieve it.

Deepening the Strategic Advantage

You should always be looking for ways to deepen every strategic advantage you have. For example, in this case, you should look at the following possibilities:

- Agree an exclusive agreement with technology suppliers
- Develop a patentable technology for the integration process
- Agree an exclusive agreement with the pharmacies

You should also make sure you don't lose the strategic advantages you already have. For example, you should restrict access to the database, ensure it's always encrypted and have backup copies in several off-site locations.

Peculiarities

This case can be used to explain some peculiarities of strategic development. For example:

- Why vision or mission statements can lead you astray
- Why excellent companies often have bland strategies
- Why you can't believe all you read

If we had started by developing the vision and mission statements, we'd probably have ended up with a completely different result. The vision might have looked something like this:

"We will provide the best optometric services in the country with more qualified optometrists and dispensing opticians than any other organization."

Any strategy built on that vision will take you in the wrong direction because you'll still be using generic processes. The problem is that if you start with a vision or mission statement, you'll tend to use the status quo as the starting point and then look for ways to make it bigger and better. When you start looking for possible threats, you're forced to take a different and more realistic approach.

Bland Strategies

Why do some leading companies have bland strategies? The answer is obvious. Would you be willing to broadcast your *real* strategy to the world? If your competitors learn you're developing a new technology, they'll do the same. You need as much time as possible to exploit your advantage, so you want to keep them in the dark. If you find yourself in a position where you must talk about your strategy, you'll do so only in the vaguest terms. You'll talk about "synergies" and "strategic deployments" but you'll never mention anything about your "StrongEye technology".

We've seen that most companies *don't* have good strategies. It's possible that these statements are a contributing factor. Average CEOs look at bland statements from high fliers and assume that's what strategy is all about!

IKEA

No chapter on strategy would be complete without a mention of IKEA. Let's look at what it does to separate itself from the competition. Some things are obvious:

- It sells furniture in giant stores located in the suburbs
- The prices are good, *and* the furniture is stylish
- The furniture is displayed in lifelike situations
- It's supplied in a ready-to-assemble flat-packs
- There's a large selection in stock
- It has a supervised play area for kids

Some things may not be as obvious:

- It designs its own furniture
- It contracts out manufacture
- It manages its own worldwide logistic system

We also need to look at what it doesn't do:

- It won't customise the furniture
- It won't provide the specific colour you want
  (if it's not already available)
- It won't assemble the furniture for you
- It doesn't provide a high level of customer service

IKEA is very profitable and has virtually no competition in its niche. It targets young and fashionable people as its main consumers – these people are often first-time home-owners who are very price-conscious. They're willing to collect the flat-packs from the warehouse and assemble it themselves, so they can afford good quality, stylish furniture.

The reason the company is so successful is that it concentrates on the needs of a very specific group and provides a highly-focused solution. Each of its activities supports the others – it's similar to Wal-Mart in that regard. If a competitor copied one of its activities (like designing its own furniture) it still couldn't compete because it needs the other elements to succeed – and the entire combination isn't easy to copy. So does IKEA have a great strategy?

- It's chosen a unique position
- It's said "no" to many opportunities
- It's designed all its activities to reinforce each other

The answer must be "yes"!

Unhappy Customers

Imagine a middle-aged customer walking around an IKEA store, searching for someone to explain how wall units can be slotted together and how to get a different colour bookcase. Finally, he leaves frustrated, and writes a stinging letter to the CEO. No doubt he gets a polite reply but do you think the operation is going to be changed, or anyone reprimanded, because of his feedback?

Of course not. He's *not* the target customer. The IKEA experience was never designed to satisfy his requirements. Remember the question at the start of the chapter? That's the answer. If you have a great strategy, you're *more* likely to disappoint some of your customers – more so than if you haven't. You'll disappoint the ones that don't fall into your target group. IKEA won't stop middle-age people coming into the shop, and many of them will be happy with their experience, but it won't make any changes to its policies or procedure if they're not. It'll react differently if the target group is unhappy.

Support

Not everyone understands this. Consider the CEO of a large company supplying two software packages based on the same code. The first package is free and unsupported. The second is supported with a guaranteed four-hour response time (Gold support). Gold customers pay a significant premium for support.

The CEO receives an email in the middle of the night. It's a user castigating him about the quality and usability of the software. He reacts immediately, calling the VP of global support and insisting that they pull out all the stops to help her. The VP calls the regional managers and they call their team leaders. Soon a crack team of support staff respond to the user. The problem is solved within the hour.

It sounds great – the stuff of PR releases and newspaper articles. The problem is, the user was using the free, unsupported copy of the software. She wasn't paying for support. That's not something you'd want to publicize because paying customers will get very upset. The problem doesn't stop there. Everyone in the support department has heard about the incident so now there's confusion: "What happens if we get a call from users of the free version?" "Do we respond to everyone now?" Team leaders and managers don't have any answers. They divert resources away from real customers, just in case the policy has changed. Strategy means little if the CEO and senior managers don't understand and support it.

Conclusion

We talked about some real companies and their strategies in this chapter. The details were based on articles, case studies and interviews. By now, you should have a good understanding of how these companies reacted to their environment and became successful. *Unfortunately, it's all fiction.* You've been reading stories that people tell, after the fact, to justify their company's success. They're all examples of *survivor bias*.

We've already seen how the human mind tries to make sense out of chaos. How did IKEA and Wal-Mart become successful? There were hundreds of thousands of decisions, mistakes, insights, coincidences, false starts and failures that brought them to where they are. Sometimes they followed a strategy and sometimes they reacted to circumstances. Some people were brilliant – but not all the time, and others were pig-headed, but not all the time. Most were average – but, again, not all the time. In the end, it all worked out, but there's no guarantee it would work exactly the same again.

There's no way anyone could explain what *really* happened. Instead, people tell stories that seem to make sense. Of course, their brains have been hard at work erasing uncomfortable facts, actuating the positive, and taking a bit more credit than is justified. In other words, even when you listen to the most honest people in the world, all you can expect is fiction. And some people lie!

Does this mean we should forget about case studies? Of course not. We need stories, even fictional ones – it's how we make sense of the world. However, *we shouldn't be tricked into believing they're true.* Think of them as parables. They offer us possibilities but they don't prove anything. Don't do something just because it's supposed to have worked for Intel, Google or Ford – do it because *you've* established that it's going to work for you.

Knowledge

You've read the case studies in this chapter and they've already diffused into your memory. You've come across thousands more – in the press, on TV, in magazines and from talking to people. They're also in your memory, perhaps mixed up with some of your own experiences. That's your core knowledge and, unfortunately, most of it is false. When you make a gut decision, you're basing it on that knowledge, and it *doesn't* provide much of a foundation. That's why you should use fact-based decision-making techniques when you're creating a strategy for your organization. They'll give you the best chance of success – and you can't get better than that.

FURTHER READING

Rumelt, Richard, *Good Strategy/Bad Strategy: The Difference and Why It Matters*, Profile Books.
Harvard Business Review (2011-02-08), *HBR's 10 Must Reads on Strategy* (featured the article *"What Is Strategy?"* by Michael E. Porter), Harvard Business Review Press.

19 MARKETING

A 2012 study by the Fournaise Group found that 80% of CEOs don't trust their marketing team. They believe marketing managers and their staff don't understand commercial realities like return on investment and creating customer demand – instead, they focus on nebulous metrics (like "followers" and "likes"). They also waste time exploring new technologies, like marketing automation, rather than doing their core job. Do you share those reservations? Remember, if they are valid, then the person responsible is you. Part of the problem may be that CEOs don't fully understand what the marketing group should be doing. We'll explore some of the key areas in this chapter.

Marketing Mix

Let's start with the marketing mix. There are four areas to consider:

1. Product
2. Price
3. Place
4. Promotion

(1) Product

This relates to the product or service you're going to supply. It'll often be influenced by strategy. Remember IKEA from the last chapter? Their products – stylish, low-cost furniture – were an integral element of their strategy. On the other hand, the products that Wal-Mart supplied were generic – they were not a key element of the strategy.

You probably don't have too much choice about the class of product or service you supply. If you run a chain of beauty salons, you're not going to manufacture jet engines. However, your strategy might define the specific product or service. If your target customer is cost-conscious, you'll probably produce a no-frills product. If they focus on prestige, you'll produce a more elaborate one.

For example, it's possible to buy a reasonable smartphone for less than €100. However, many people pay up to €1000 for the latest Samsung or Apple device. You'd expect there would be significant differences to justify that price difference. There are some differences – at the top end, you get better materials, faster processors and higher resolution cameras. However, the core functionality is very similar, so on a purely rational basis you couldn't justify those high prices – but people do. That's something we've seen before. People are not perfectly rational – System 1 makes the decision and System 2 is pulled in to justify it. They want to have the latest device like their friends, so they find a reason: "It's got a 35 Mpixel camera, 3.3 MHz processor and 650 PPI screen."

They've made their decision for emotional reasons, but they justify it using logic. Their new phone is undoubtedly better than their last, but they'll never notice the difference in ordinary use.

So what factors should you consider when you're about to develop a new product? Here are some key ones that should be included:

- The customer
- Your brand
- The competition
- Your existing products
- Your capabilities

The Customer

One of the goals of marketing is to ensure that your products or services fit the needs of the target customers. So, before you create a new product, you should find out what customers need, and what they're willing to buy. There are many techniques used to do that:

- Surveys
- Interviews
- Focus groups
- Product testing

This is called market research and it should eliminate guesswork and "gut feel", but results suggest otherwise. Estimates vary, but it's likely that at least 80% of new products fail. That's a shocking statistic and it indicates there's something seriously wrong with the market research process that most companies follow.

One reason is that customers don't make decisions in a rational way. When you ask someone how they feel about a new product, their System 2 kicks in and they'll give you a rational answer. However, when they go shopping, System 1 is in charge. Afterwards, if you ask them why they made that selection, they'll give another rational answer. It's what they believe, but it's not the truth.

Savvy start-ups are aware of this and use a different approach. They build prototypes and then actually sell them to a small section of their target market. If people aren't buying or there are problems with the product or service, they re-design it and try again. This has the advantage that the response variable (number of units sold) is grounded in reality – there aren't any over-optimistic assumptions. However, it can mean putting a sub-standard product onto the market that could tarnish a big brand name. It might also necessitate many iterations to get to an acceptable offering.

Marketing people spend lots of time creating customer profiles. These can contain lots of detail that have nothing to do with the product or service you provide. That's because conventional wisdom states that the more you know about your customer, the better your predictions will be about whether they'll buy your product or services. But is that true? If it was, then you'd already know what products your friends are going to buy (without asking them or looking in their cupboards).

Many core beliefs in marketing (as in other areas) are based on gut feel. They seem plausible, so they're accepted without question. As we've seen, that's a dangerous and wasteful approach.

It's your job to make sure that any practices that consume time or money are evidence-based. They should be probed and validated before they're used to make decisions.

The Brand

When Apple develops a new smartphone, people get excited and there are launch-day queues. The press is full of reviews and people can't wait to get the device. If another company produced the same device, they'd get a more muted response. Why is that? Part of the answer is the Apple "brand". They've created an intangible asset that's probably worth as much as their physical ones.

You could say a brand is just a name, a logo, some text and a selection of colours, but that's simplistic. Some companies have turned their brand into a strategic advantage. What happens when you hear the name "Ferrari"? You might become aware of some of the following impressions:

- Fast
- Red
- Sports car
- Race
- Rich
- Glamour
- Excitement

That's your associative memory (System 1) in action. Each impression is linked to the next. You've been conditioned to activate this cluster when the name "Ferrari" is used. But can you remember the logo and its background colour? Most people know the name but not the logo. If you're a car enthusiast, you might recall the prancing horse. Even then, it's unlikely you'll remember the background colour (yellow) or the three other colours – the green, white and red of the Italian flag.

So a brand is not a logo or a few colours. You don't spend €276,000 on a car just because of the logo. The brand is the feeling that people get when they see or hear about the company's products. It's the ideas and emotions that are unconsciously triggered.

Why do people buy a Ferrari? It's not just because it's an amazing machine (that's System 2 justification). The primary reason is an emotional one. If you buy a Ferrari, you're hoping all that glamour and excitement will rub off on you. You're hoping that other people – particularly your peers – will envy you.

Rebranding

Marketing people often spend millions on changing names, colours and logos, but do those changes make a significant difference? Guinness had declining sales until it was relaunched in 1981. Despite changes to the product, its logo (the harp) and colour scheme were retained – and it was a success.

Volkswagen held on to the Škoda name and retained the logo (the winged arrow), despite its negative reputation. That didn't stop it becoming a success. Of course, these examples don't prove that a logo change *won't* help sales, but many companies have wasted money on such exercises. You need to be sure before deciding on a "rebrand".

Before you create a new product, you need to analyse how it fits with your brand. For example, if Apple produced a plastic €59 iPhone, its brand name would probably help it sell a lot of units. That's called "brand extension". You get more sales because of your reputation in a related niche. Is that a good idea?

Write down your thoughts before reading on (Q19-1).

It seems reasonable to get the most from your assets, but studies have shown that it could dilute the brand image. That's the opposite of a good strategy. Would you *really* feel the same about the Apple brand if it includes low-end throwaway phones as well as more expensive ones?

The Competition

If you have a profitable business, it's almost certain you have competitors and your success is inversely linked to theirs. Before you create a new product or service, you should determine where it fits in relation to competitor products. Is it addressing a gap they haven't seen? Is it going head-to-head with products already on the market? Does it offer better value for money or a lower price than the competition?

You'll also need to consider what competitors will do when you launch your product and estimate the probabilities of the various responses.

For example, how likely are they to:

- Drop prices on competing products?
- Drop prices on all products?
- Launch a competing product?
- Launch a product to compete with your cash cow?
- Offer a package deal?
- Increase their advertising?

You'll need to decide what you'll do if any of those things happen, and what *their* response is then likely to be to that. This is a form of *game theory*. It forces you to look several steps ahead and consider likely outcomes in advance. If that makes your head hurt, you know the reason – System 2 is trying to avoid the effort. You need to work with your marketing team to make sure they've done a detailed analysis, so you can avoid any pitfalls.

Existing Products

There's little point in having two or three products with the same features. You're likely to confuse customers and you're also likely to waste money advertising and promoting them. It's best to have a well-differentiated spot for each product. That might mean cutting back on existing products and being careful where you position new ones. You should also consider the product lifecycle. Is it possible that the price will change over the new product's life? (Don't guess – look at the data.) If that happens, will it conflict with any of your older products?

Capabilities

You know more about your organization's capabilities than anyone else. Would you agree?

Write down your answer before reading on (Q19-2).

That statement sounds reasonable and many CEOs will instantly agree – but it's almost certainly not true. In the first place, it's unlikely that you, personally, know more than *everyone* in your organization. Much of the knowledge is in the brains of workers, technical people, team leaders and managers. But even if you combine all that knowledge with yours, it still mightn't be sufficient. Remember, outside the organization, there are vendors, process experts, researchers and thousands of others who may be working on similar products, equipment and materials in ways you can't even imagine.

The point is that it's easy to become short-sighted and assume you're the guru in your chosen area. Just because you can't imagine how something could be improved doesn't mean other people can't. When you're considering a new product, you need to expand your horizons. Talk to people inside and outside the organization and be open to new ideas, no matter how offbeat they seem.

(2) Price

The second element in the marketing mix is price. We've already looked at many of the factors you need to consider in this area. Now we're going to look at the relationship between features and pricing and their effect on positioning. Have a look at the following graph. It shows a relationship in the smartwatch market:

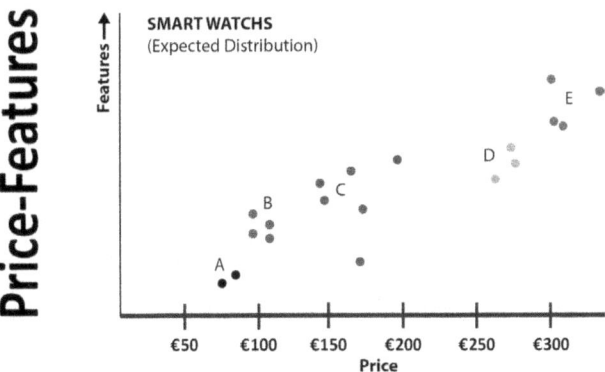

The X-axis shows prices and the Y-axis shows features. There are no smartwatches under €70. There are two in cluster A between €70 and €90. There are four in cluster B, clustered around €100. Cluster C is spread out between €140 and €199. Then we have cluster D (€260-€270) and cluster E (€280-€350).

Generally, features increase as the price goes up – as you'd expect. Review the information for a few minutes, then assume you're about to create a new smartwatch and write down your answers to the following:

1. What's the best price to set?
2. Is the price-features relationship as you'd expect?

Write down your answers before reading on (Q19-3).

It looks like there's a gap in the market between €200 and €260. You might think this is the perfect place to set a price for your new smartwatch, but you might be disappointed – it could be a natural "break point". In other words, one group of customers is willing to pay up to €199 for a smartwatch – but no more. The next group is willing to pay more but is brand conscious. There's a small clue in the data to indicate this is the case – cluster D watches have similar features to the high end of cluster Cs but are much more expensive. This means some customers are willing to pay considerably more for the same set of features – an indication that intangible factors are at play. This example demonstrates that setting a selling price is not as simple as looking for a gap. Just because you find one doesn't mean there's a real opportunity.

Complications

The previous graph makes logical sense – it's what you'd expect to happen. Unfortunately, it's not real (did your System 1 accept it without question?). The real smartwatch market is much more complicated and confusing. The following graph gives a better indication of what it really looks like (but still doesn't capture the real story!):

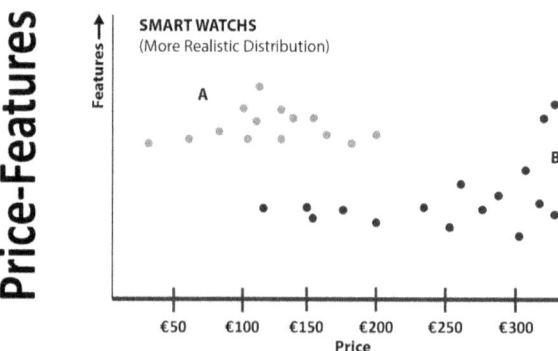

There are two main clusters. Cluster A on the top left offers watches that are also stand-alone phones. They can make and receive calls and can also link with smartphones. They have Bluetooth and Wi-Fi. Many of them have GPS and heart rate monitors. Some even have cameras. Cluster B consists, primarily, of smartwatches that only link to smartphones. All have Bluetooth, and some have GPS and heart rate monitors but, generally, they have fewer features than the first cluster.

How can that happen? You might think it's due to the quality of the watches but many in the first cluster have high-quality components. There *is* a branding issue. Most of the watches in cluster A are from Chinese companies that are not well known in the West. However, there are also several watches in cluster B that come from relatively unknown companies.

There are many factors that feed into the price-feature equation. Customers want different things from a smartwatch. Some want every feature possible, some want a small subset of features, others want a device that'll impress people. Before you set a price, you need to understand the needs and expectations of the various groups and decide how you're going to address them. However, even within a sub-group, it doesn't follow that you'll get a higher price by adding features.

This is (again) where data beats gut feel. Like everyone, marketing and sales people make assumptions. They believe they know how things work, but that's their System 1s in action. Their explanations may seem logical (like the first graph above) but you need data to confirm you're getting the real picture.

(3) Place

The third element in the mix refers to where the customer and your products and services interact. Consumers once went to speciality shops for everything. They got bread at the bakery and tools in the hardware store. If they wanted a suit they went to the drapers. Now there are big-box stores that sell everything in one place. That was a profound shift for customers, retailers and suppliers, but it shows that expectations can be broken.

Where do *you* sell your product? Do you use retail outlets, or do you ship directly to the end customers? Could you do it differently? Consider how the Internet has changes things over the last few years:

- Books are now sold and delivered over it
- Music and movies are streamed over it
- Banking services are provided over it

If your product has a large information component, then it's the perfect way to both sell and deliver to your customers. But it's not the only technology that'll affect where goods are sold, delivered and consumed. There are more changes coming. For example:

- 3D printing
- Drones
- Driverless cars, vans and trucks
- Robots
- Augmented reality
- Blockchain and bitcoin

These and many other technologies could have a profound effect on where you'll sell and deliver your offerings. You need to look ahead and be ready to change.

Services

If your organization only provides a service, you might think you're immune from technology change. That's not necessarily true. For example, one researcher estimates that between 30% and 50% of the work being done by car mechanics could soon be done by machines. A McKinsey report estimated that 60% of all occupations could be at least 30% automated. That level of automation could affect where your service is provided, and if your competitors use new delivery channels first, you could be left behind.

(4) Promotion

The fourth element in the marketing mix – promotion – is where the marketing department focuses most of its effort. There are five channels you can use to get your message out:

- Advertising
- Media coverage
- Personal selling
- Websites
- Social media

The first three are the traditional channels that have been used for decades. *Advertising* includes print, radio and TV advertising as well as billboards, point-of-sale messages and packaging. *Media coverage* is also called public relations. These are the stories about the company and its products that appear in the press and social media. *Personal selling* is the oldest sales method of all and is still used for more expensive items, but companies like Tupperware and Amway have also created a clever variation for low-cost goods. Telemarketers sell goods and services in the medium price bracket.

Websites and *social media* are relatively new but have already made a big impact on the promotion element of the marketing mix. Of course, traditional advertising is also used on the web and can be particularly effective when used on search engines (e.g. AdWords) and social media platforms (e.g. Facebook).

Social media adds a new dynamic to the marketing equation because customers aren't just passive recipients of your message any more. Consumers can let everyone know if your product meets their expectations. Potential customers will search online for reviews. They'll watch your product being demonstrated by customers on YouTube and read the reviews on Amazon. They'll put more faith in those channels because these are less likely to be misleading – unlike your carefully crafted messages. Research has confirmed that after family and friends, social media channels are now the most trusted sources of product and service information.

Love Your Haters

We saw earlier that if you want to maintain a strong strategic advantage you'll have to upset some people. What does that mean for your social media exposure? It means you'll get people who'll publicly complain about your organization and its products – the technical term is *haters*. That's going to upset your marketing people. They want a nice clean positive message out there – not people who are shouting about the negatives. The natural reaction is to either solve their problem or shut them down. However, both tactics could be counterproductive. Can you see why?

Write down your answer before reading on (Q19-4).

If someone complains that your products don't have a specific feature, you might promise to put it in the next version. However, if it's not important to your target customer group, you're just adding cost for no gain. You might even alienate some of your core customers. You could try to argue that no sensible person would want such a thing, but that approach could backfire and make things worse.

In fact, it can be an advantage to have some haters on social media because their comments can trigger others to defend your brand. These are called *advocates* and can have a powerful influence on people in their circle. They are often better than traditional channels for getting your message out – and their voice will be more convincing.

Your marketing manager should know who the most important influencers in your industry are, what they're interested in, and what they're currently talking about. He should communicate with them, so *they* get interesting information for their followers and *he* gets useful feedback. Has your marketing manager been developing those relationships?

Market Segmentation

You could try selling to everyone, but it's more powerful if you can concentrate on a specific niche. You're more likely to build a strategic advantage (like IKEA) when you have a well-defined target group of customers. We've seen that people have different needs and interests. A newly married couple will have different priorities to a couple approaching retirement. A single professional woman will have different interests to a middle-aged mother. It makes sense to concentrate on segments that best match the brand's objectives.

Segmentation requires a choice. It means saying "no" to many opportunities. It's an important decision and needs to be undertaken with the same level of deliberation and using the same tools we've discussed earlier. Once you've made that decision, you can use all five promotion channels to deliver a consistent and carefully crafted message to the right people. If your offering is unique, it meets the needs of your target audience and your promotions do it justice, then you're more likely to succeed.

Advertising and Data

You can spend millions on advertising. That's fine if it works, but do you know whether it does or not? Billions have been wasted on promotions over the years and one reason is that many CEOs are unwilling to address that core question – so let's do it now.

Let's start with the first type of advertising that used reliable quantitative methods: direct mail. The concept was beautifully simple. You put an advertisement in a newspaper offering a special deal. If people were interested, they'd send in their money and get the goods in the post. It was easy to find out if a particular advertisement worked. You counted the number of responses, multiplied by the amount each paid, and you had your answer.

If the ad worked, you could just put it in again – but you might be able to do better. You could change a few words or the picture to see if you got a better response. But that raised a problem – if you got a better response, was that due to the changes or did it just happen by chance? Perhaps people were more inclined to buy your product the second time because of the weather or some other external factor?

That's when you'd use A vs B testing. You split up the country into two areas, each having roughly the same characteristics – you proved that assumption by sending out the same ad to each area at the same time. You should get roughly the same results from each. Next, you sent out your new ad to one area and your old one to the other at the same time. If you got significantly better results, you knew that the new one was better. Over the years, direct mail professionals got very sophisticated because they tested everything – there was no need to make assumptions. If someone said that a bigger picture was better, that hypothesis could be tested – and the results would decide.

This level of clarity never reached other forms of advertising. If you put a quarter-page ad in a national newspaper, how could you be sure that any apparent change in sales was due to the ad? Most organizations didn't even try to quantify results. You've heard the expression: "I know that half the money I spend on advertising is wasted. The problem is, I don't know which half." The trouble is, it may not just be half – it might be 80%, 90% or even 100%!

The Web

Finally, the world-wide-web arrived, and people seized the opportunity to get accurate data. Nowadays, almost everything can be measured. You can use analytics packages to track:

- Total number of visits to your website
- Traffic sources
- Bounce rates
- Best performing pages
- Conversion rates

You can track visitors as they move between pages on your website and you can see where they leave your site. You can even perform dynamic A vs B testing to make sure every element of your message has been optimized. Do we finally have the right data for reliable decisions?

Maybe not, because those results are not as dependable as they seem. In 2016, Imperva, a US cyber security company, analysed web traffic and estimated that 52% of it was due to bots. These are small programs that carry out automated tasks on the web. They can visit your website and act like a human visitor. They can move from one page to the next – they may even try to buy something from your site and follow it through the checkout process.

So, even though you can be very precise about the number of visitors to your site, you can't be sure how many are human. The best guess – and it's only a guess – is that slightly less than 50% are. Why does that matter? Because your staff could be spending hours optimizing messages, fonts, colours and pictures based on data generated by bots – who don't really care. They're just carrying out a set of pre-programmed tasks and aren't influenced by any fine-tuning you do. To make things worse, if you're paying for advertising on the web, you may be wasting even more money. It doesn't matter whether you're paying per click or per impression, it's possible (and likely) that some of your hits are caused by bots. You're paying for the privilege of showing your message to a piece of code.

But all is not lost. Your technical people can set up filters that will include only valid hostnames in the data and there are resources that track bot activity and provide lists of suspect addresses. Nevertheless, it's not as easy as it should be to get accurate results and you can never be certain about your traffic's integrity – the bad guys are innovating all the time.

Probabilities

We've seen that you can't really be certain about anything in your external environment. Your competitors may react in a certain way, your customers may react in a different way and some of the data you're using to measure your performance is probably misleading. Gut feel is not a solution in that type of environment. You need to harness System 2. You need your marketing team to work with data, evaluate alternatives, and assign probabilities. You have one dependable response variable – your sales revenue. That metric can help you identify what's really working and what isn't.

Review

So, is *your* marketing working? If it is, you should have answers for the following questions:

- What category are you competing in?
- Who *are* your target customers?
- Why do they buy from *you*, and not your competitors?
- Who are *not* your target customers?

That's a good start, but you really need to know much more.

For example:

- Who are your competitors?
- What are they doing that you can't match?
- What is your unique selling proposition (USP)?
- Why can't your competitors duplicate your USP?
- Does your brand have a unique identity?
- Does your advertising reinforce your brand identity?

If you don't already know the answers, then you and your marketing manager have some work to do!

FURTHER READING

Kotler, Philip; Kartajaya, Hermawan; Setiawan, Iwan; , *Marketing 4.0: Moving from Traditional to Digital*, Wiley.

Ries, Al; Ries, Laura, *The 22 Immutable Laws of Branding: How to Build a Product or Service into a World-Class Brand*, HarperCollins.

20 TRAINING

Any senior manager will tell you how important training is in their organization. When prompted, they'll produce a training matrix that shows the courses people have taken and tell you exactly how much money has been spent over the last year. But ask them how effective the training has been, and you'll get platitudes: "It's been very important to the development of the organization." "It's the key to our success." Is it really? When you look closely at their training programmes, you'll probably find that:

- Staff are told what courses they must take
- There's no preparation before the course
- The course is delivered in one block
  (e.g. a three-day course is run Tuesday to Thursday)
- There's no evaluation of training outcomes
- There's no evaluation of behaviour changes

If that describes your training programmes, then you're not just throwing money away, you're also throwing away the opportunity to improve your organization. That sounds terrible, but it's actually good news. First, let's look at the reasons why training doesn't work.

Memory

Humans have bad memories – really bad. Let's assume that a trainee on one of your courses is exposed to 100 facts. Let's also assume (for the moment) that at the end of the course they can remember all 100 facts. Once they leave the room, they immediately start to forget. After the first hour, *they've already lost 60 facts*. After 24 hours, they can only remember 28, and by the end of a month, that's dropped to 15! This incredible memory loss is illustrated by the "forgetting curve" (there's a version on the next page).

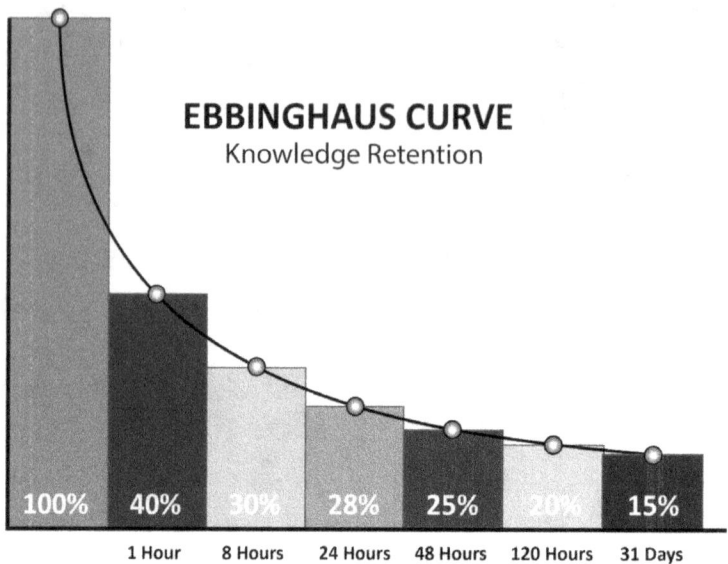

It was discovered by a German psychologist called Hermann Ebbinghaus at the start of the 20th century and has been repeatedly confirmed by researchers since then. *There's no doubt that it happens.*

This curve is an average. Some people do better, while others do far worse! Before we look at the implications of that memory loss, let's go back and examine a very shaky assumption we made earlier. We said that if a trainee had been exposed to 100 facts she'll remember all of them at the end of the course. *That won't happen.*

Working Memory

Before you can remember a piece of information, you have to capture it using your senses (hearing and sight, primarily) and hold it in "working" (or short-term) memory. It must stay there long enough to be encoded into long-term memory, and that could take many minutes, even hours. The trouble is, researchers now believe that *working memory can only hold four chunks of information at one time.* Can you see the problem? That's right, if the trainer is throwing out lots of facts, any items going into working memory are immediately displaced before they have a chance to be encoded – they never get a chance to be laid down in long-term memory. If a trainee is exposed to 100 facts in a training session it's likely that, at most, he'll remember 10 or 15 at the end of the session.

And, as the number of facts increases, retention gets poorer. So, things are much worse than we estimated in the last section – you'll be lucky if any trainee remembers four or five facts by the end of the month.

Attention!

We've also assumed that the trainee was *concentrating*. However, the longer the session lasts, the less likely that's going to be true. It's estimated that, for most people, *20 minutes is the attention limit*. After that, concentration levels drop, and minds move on to other things. Most of what the trainer is saying is filtered out like an annoying buzz at the edge of consciousness. Continuous training sessions of an hour or more become pleasant daydreaming interludes.

The Subject

There are other factors that can torpedo your training course. For example, if trainees have no interest in the subject matter, they'll filter out the material long before 20 minutes is up. And, if the material is presented at the wrong level – too low or high – they'll either get bored and stop listening – or become bewildered and shut down. We could also talk about the *structure* of the information or the *delivery style* of the trainer and their ability to react to feedback. These can further reduce the amount of information that's retained.

What Do You Want?

One of the problems is terminology. People talk about "giving training" when they should be talking about "learning". A trainer can *give* a great training course, but nobody might have *learned* anything. We want people to learn, but that not enough either. We need to take things one step further. The ultimate purpose of training is to produce *a behaviour change*. It might be to perform a new task or to improve performance on an existing task, but *if there's no behaviour change, then the training hasn't been successful*. People will argue the trainees have gained useful foundation knowledge, but what's the point of gaining knowledge if it doesn't result in a change? Besides, knowledge will quickly fade away unless it's used.

Measurement

How do you measure training? Do you use feedback sheets? That's an excellent example of a *proxy* measurement. You're measuring how *popular* the training is, not how *useful* it is. You're also motivating trainers to provide an enjoyable experience and avoid anything that could cause discomfort – like forcing attendees to use their System 2. Many trainers have learned to game the system by providing *entertainment* instead of *knowledge* and *expertise*.

The Right Questions

So, before you authorize a *learning* course, you need to be sure that adequate preparation has been done. You need solid answers to the following questions:

- What are people doing *now*?
- What *behaviour change* do we want to achieve?
- What is the benefit for the organization?
- What is the *least* amount of information needed?
- How will the information be structured?
- How will the learning be structured?

Then you can get into the implementation details:

- How can that information be captured and displayed?
- Who can present that information properly, and how?
- Why will people *want* to learn this?
- How can we make sure it will be *used*?

Finally, you can look at the outcome:

- How will the changes be implemented, and when?
- How will it be measured?
- How can we continue to improve the learning process?

When you're *sure* a course makes sense, you must appoint someone to do the work. They'll solicit input from participants *before* the course, and agree specific *outcomes*. They'll create the learning materials and integrate theory and practice in a way that'll make sense to the learners. They'll also optimize the scheduling, train the tutors in the new methodology, and much more. Yes, it takes a lot to develop a learning system that works, but the alternative is wasted time, effort and money.

The Cost

When a learning system *doesn't* work, the cost to your business is high, but it's often hidden. It manifests itself when jobs aren't done properly. It's the reason why tasks take longer than they should, and why you have quality problems. It's why there's uncertainty, confusion and a never-ending cycle of problems that crop up, get "solved" and reappear again.

Back to You

You already know that training doesn't work. To confirm it, all you have to do is recall one of the courses *you've* attended and answer two simple questions: "What do I remember about the content?" and "What behaviour has changed since the course?" If you *still* have doubts, you can carry out another check. Ask *anyone* who's attended a course in the last three months to write down what they remember about the subject matter. It's easy to tell whether training is working or not.

Good News

If your training courses are *not* working, that's good news. Why? Because you have a unique opportunity to make significant gains. If you can put an effective *learning* programme in place, the medium- and long-term benefits will be substantial.

21 THE FUTURE

Nobody can accurately predict what's going to happen in the future – but there are certain trends that already look like they're going to have a *significant* impact on our world. We're going to look at several that could directly affect *your* business and your job.

Moore's Law

You're probably heard of *Moore's Law*. It stated that the number of transistors in an integrated circuit doubles every year (later changed to every two years). That doesn't sound particularly exciting, but it's incredible because it's a description of exponential growth. In 1972 the transistor count in a microprocessor (the Intel 8008) was 3,500. That was a great achievement and ushered in the era of the personal computer. Now, there are processors with 10 *billion* transistors! Even the processor in a smartphone – like the iPhone 6 – has over 2 billion transistors.

And it's not just processing power that's increasing – the size of memory and storage has also increased, and prices have fallen dramatically. It's common for home users to have several terabytes of data. Business can have hundreds of petabytes available, so storage is no longer a limiting factor, particularly with the availability of low-cost cloud storage.

The Smartphone

It's not just a matter of cramming more processing power into desktop computers. The *smartphone* has drastically changed how we interact with technology. We now take our processing power with us and we're in constant contact with servers around the world. We're hooked up to a range of sensors, including cameras, microphones, gyroscopes, accelerometers, proximity sensors and much more, and they're active all the time. GPS shows us *where* we are (so we never get lost) but it also talks to other apps behind the scenes and they use that information to provide additional services – like telling us about that new restaurant or calling a taxi to our location. Your smartphone is your personal interface to a host of technologies. It can instantly become a drone controller, a home security monitor or a smart thermostat controller. Now, consider everything it can do today and realise that's just scratching the surface because there are thousands of developers dreaming up new applications every week.

Tools

It's not just hardware improving at an accelerated rate – the same is happening with software. That's because people aren't just creating programs, they're also creating the *tools* to create them. Programmers can build on existing platforms and pull in libraries of code that have already been developed. They don't have to start from scratch each time. That means it takes less time to develop new programs and each can be more complex than the last. They can develop programs in days that would have been *impossible* less than a decade ago. And the rate of change is constantly increasing.

Learning From History

We think about the future in terms of what already exists because we use our knowledge and experience as an anchor. When mobile phones were introduced, they were simply a mobile version of the office telephone. Few people could have foretold how they were going to evolve and change our lives.

However, we're approaching a time when that level of understanding will be difficult because change will be coming too fast. We won't have time to become comfortable with the current wave of innovation before the next one hits us. Many of these changes will have political, cultural and business implications that will threaten the stability of institutions and may even change our entire belief system. You might argue that if that *does* happen, we can simply slow down or allow the wave to pass us by – but that may not be possible. For example, if an innovation creates a substitute for your product or service at half the cost, can you afford to ignore it? And, just as you're reacting to that challenge, if another one comes along that halves the cost again, can you ignore *that*? The future is not going to be like the past. Things are going to change much faster and in ways that we can't imagine.

The Internet

To see how quickly change can happen, let's look at the Internet. It only became available to the public in 1989 and for the first few years, there were probably less than a million users. That increased to 16 million by 1995. Yet just 22 years later, in 2017, it was estimated there were 3.9 *billion* users – that's 47% of the world's population. In developed countries, the figure is over 80% of the population. The Internet now provides a platform for millions of businesses – from Google and Facebook to one-person start-ups. Most of these could never exist without it. Many of the biggest businesses in the world are operating in an environment that didn't exist 35 years ago!

The next innovation in this area could be the *Internet of Things* (IOT). Every electrical or electronic device will have an IP address and be connected to the web. Your fridge will know when it's out of milk and will automatically order from the local grocery shop. It will be delivered by a drone that will drop it gently on your rooftop landing pad. Your door will detect if someone calls while you're out and ask them to leave a voice message. Your toaster will know the moisture content of the bread (based on the age of the bread and the humidity in your house) and will heat it accordingly. Technology will be embedded all around us.

Robotics

Robots have always been the poster children of the future. They're widely used in industry but have made limited inroads in small companies and our homes – vacuum cleaners and lawnmowers are the exception. Now there's a new breed of low-cost human-friendly robot on the way. These don't have to be programmed – instead, they can just be shown what to do. They'll also be safe working next to humans. That's important because most industrial robots are confined to cages – they're too powerful to work directly with people. As prices come down, the new robots will become more common. You might see them stocking shelves in the supermarket or bagging your groceries. They may also be able to build your house, service your car and even fix your phone.

3D Printing

3D printing is the process of making a three-dimensional object like a cup or a knee joint by printing it – it's like an inkjet printer. This process is well established in industry. It's used by healthcare companies, automotive suppliers and more. It's likely that we'll have a similar system in our homes before too long. There are low-cost versions available for hobbyists right now, but these are slow and have limited capabilities. However, personal computers first became popular with hobbyists, and this technology could follow the same development path. Once 3D printing does mature, it will have a significant effect on the economy. You won't go to a shop to buy objects like crockery, cutlery, ornaments or even clothes. Instead, you'll download a digital blueprint over the Internet and print out the object.

Virtual Reality

Virtual reality (VR) already exists. Millions of people spend their leisure time playing online games. One of them, *League of Legends*, is estimated to have over 20 million daily users and there are many others, including *World of Warcraft* and *Dota 2*. Some virtual worlds like *Second Life* provide an experience that mimics real life. People buy virtual goods using virtual money and socialise with other players.

You can now experience some of these worlds with a 3D headset like Oculus Rift and soon it'll be difficult to tell if you're in the real world or a virtual one. This could have a significant effect on institutions and businesses. Take education – why send your kids to school when they could put on a headset and experience the classroom from home? Education will be very different when students can "meet" historical figures like Leonardo da Vinci and observe Galileo drop weights from the Leaning Tower of Pisa!

Most of the focus on VR is on virtual worlds but the same technology can be used in other ways. For example, a robot could be controlled by a person wearing a VR headset. When they lift their arm or turn their head, the robot would do the same. This means jobs that entail physical work, like engineering, maintenance, or security could be carried out by people from their own homes. If VR does become widespread, it could reduce travel significantly and much of our existing infrastructure – like schools, churches and offices – may not be required. It will also generate *new* opportunities, but these will reside in the virtual world.

Voice and Face Recognition

Voice recognition systems have come a long way over the last few years with applications like Siri, Google Now, Cortana and Alexa gaining widespread acceptance. When their accuracy approaches 100%, we can expect to see the technology embedded in a wide range of applications. For example, you might go online and book an airline trip without typing anything – just speak your destination, date and time of departure and a few personal details and everything is done for you. You may not even need a ticket. When you book the flights, your face and voice could be analysed. When you arrive at the airport, the system will recognise your face and confirm the identification with your voiceprint. You'll just walk through to the departure lounge.

Driverless Cars

Many companies, including Google, have demonstrated driverless cars, so it's no longer science fiction. Most car companies including Audi, Ford, BMW and Tesla expect driverless vehicles will be commercially available before 2022. It's likely that specific applications will be targeted first. *Long-haul trucks* are an obvious application, because most of the driving takes place on motorways and the interaction with other traffic is kept to a minimum. Local taxi services could be another early application. That's because taxis are restricted to a relatively small local area that can be pre-mapped in detail. Uber is one of the many companies carrying out research in this area.

Blockchain

You've probably heard of *Bitcoin* - it's digital money with no central issuing authority. That means it's not regulated by any government or central bank. However, the technology behind Bitcoin could be even more significant in our business and personal lives. This is called *block-chain* and it has the potential to change the fundamental administrative and legislative framework of our entire economy.

Remote Applications

The advantage of having every device connected to the Internet is that individual devices don't need to contain *all* the required technology. For example, your fridge could have *face recognition* to track what you're eating. The recognition software could be hosted in the cloud – it may not even be owned by the fridge manufacturer. An encrypted biometric sample will be sent to the application over the Internet and it will send back confirmation of identity. This approach will speed up the implementation of new technologies because there are huge economies of scale. The face recognition system on your fridge could also work on your front door and in your driverless car.

The Economic Fallout

There's a potential problem with all this automation. What about the people who'll lose their jobs? We've already mentioned long-haul truck and taxi drivers, but that's just the tip of the iceberg. Think about the bank clerks, call centre operators, checkout staff. Very few jobs are safe. What will happen when these people no longer have jobs – no longer have salaries coming in? Where will your customers come from? This time, it's not *just* industrial workers who'll suffer.

Intelligence

Up to now, people who worked with information were safe. That isn't true anymore. Take doctors. Much of the knowledge they use for diagnostic purposes has already been codified, and many of the big tech companies – including Google – are experimenting with artificial intelligence (AI) for diagnostic purposes. It won't be too long before these systems will offer an alternative to the traditional doctor's surgery. Lawyers, accountants and consultants will also face competition from automated systems, and they'll be more accurate and more cost effective than humans could ever be.

AI

AI means that machines can reason and make decisions on their own. Most computers don't do that – at least for the moment. Your bank probably uses a program to decide if you should get a loan. It'll work through a series of questions, including:

- What's the balance in your account?
- Have you ever missed a repayment?
- How much are you asking for?

The program will collect all the data, crunch the numbers and come back with a conclusion. However, it was a *human* who decided what questions to ask, what weighting to give them and what criteria to use for acceptance. The machine decided nothing – it just carried out the calculations.

AI is different. It looks at examples – called "training data sets" – and it's then able to decide if you're a good prospect. Nobody tells it what to check for – it makes up its own rules. It does that by looking for patterns in the data and then creating rules from them. Of course, that means it *can* make mistakes – we've seen there will be random patterns in any large data set. It might find, for example, that a large proportion of all males with brown hair and blue eyes default on their loans. That might lead to a false conclusion – one that's unfair to all brown-haired, blue-eyed males!

Despite the risks, AIs trained with large data sets are likely to achieve high levels of accuracy. And "real" AI programs are not just designed to solve a specific *type* of problem. In theory, they could be working on bank loans today and diagnosing a patient tomorrow. They're like humans in that respect.

AI Milestones

The history of AI stretches back many years, but real progress only began in the late 20th century. In 1997, Deep Blue, the chess-playing system from IBM, beat the World Chess Champion Garry Kasparov. While this was an impressive achievement, it used brute force tactics – the system could evaluate 200 million positions per second and search up to twenty moves ahead. It wasn't true AI, because the evaluation heuristics were coded by humans.

In 2004, Professor Levy from MIT and Professor Murnane from Harvard listed professions that were most or least likely to be auto-mated. They felt that truck driving couldn't possibly be automated in the foreseeable future. We've seen they've already been proven wrong. Technological progress is moving so quickly that even experts can't keep up.

In 2011, a computer system called Watson defeated two former champions on Jeopardy, a US general knowledge quiz. This was im-portant because the system had to "understand" the questions and then find an answer that would match the quirky format.

In March 2016, a Google system, DeepMind's AlphaGo, defeated go champion Lee Sedol. Go is significantly more complex than chess and requires intuitive, creative and strategic thinking. AlphaGo used neural networks to win. The strategies and tactics it used were not hard-coded – the system created them by "watching" millions of games and by playing itself. In effect, it taught itself what to do.

There are many indicators that AI is going to be important. In 2017, Microsoft mentioned AI in its vision statement: "Our strategy is to build best-in-class platforms and productivity services for an intelligent cloud and an intelligent edge infused with artificial intelligence." In the same year, the UAE appointed its first Minister for Artificial Intelligence.

Solutions

So AI is making rapid progress – but how could it affect you? Let's consider an area where technology has already made inroads. Imagine a CEO in a company with four plants around the world. She wants to attend the weekly management meeting in each plant but doesn't want to travel, so she uses a telepresence system. She can see her managers in real time on her computer and they see her image on a large monitor in the meeting room. When this is set up properly, it almost looks like she's actually present. This technology is common today.

Now imagine she's got a bad cold and has to stay in bed. She still wants to attend the meetings, so she gets an avatar to stand in for her. This is an animated picture of her face and upper body. The head turns, the eyes blink and the mouth moves in synchronisation with her voice. It looks and acts exactly as if she's present. This is *also* possible with current technology. Now let's take the next step.

She's not available. Instead, an AI system, with face- and voice-recognition and a speech synthesizer, is connected to her avatar. The system has been trained with transcripts and videos from hundreds of company meetings and thousands of generic meetings. It can ask questions, make comments and even make decisions based on the input it receives. How could you tell the difference?

Decision Making in the 21st Century

It sounds like science fiction – and there's still some way to go – but the foundations are already in place. A Hong Kong company called Deep Knowledge Ventures appointed an algorithm called VITAL to their board with voting rights back in 2014. So the idea that a piece of software can be granted decision-making power has already been accepted in principle.

The Singularity

And that brings us to the *Singularity*. This is the prediction that once artificial intelligence has reached a certain level, it will trigger runaway technological growth. That could result in fundamental changes to human civilization. It's easy to imagine how this might happen. A relatively simple AI system is given the task of improving its own program. It succeeds in making it a little better, and then hands the task over to the new version. This program also makes improvements and hands it over to next generation. This continues, with each generation getting better than the last.

However, this process could happen blindingly fast, perhaps thousands of times per second, and the cumulative effect of these improvements would be immense (remember exponential growth?). It could result in an intelligent machine with capabilities that we can't imagine. It would know more than any human, it could make better decisions faster and it wouldn't suffer from the biases and inconsistencies that have plagued human decision making.

Human CEOs

There's little doubt that machines will start to improve their own algorithms and become smarter. Sooner or later, the first business owner will hand over the running of their business to an AI entity. And sooner or later, the AI will be successful because it will base its decisions on *evidence* and *probability* – not System 1 and *emotion*. That'll be the beginning of the end for the human CEO.

That's such a breath-taking conclusion, you might be tempted to dismiss it. You might argue that a piece of software could *never* achieve better results than a human with years of management experience. It just wouldn't have the practical knowledge to make good decisions. But let's think about that. Let's assume a CEO is considering taking over a company. The human CEO will get her managers and staff to do the homework. She'll review *some* of the information and talk to *some* of the people, but she'll still spend most of her time on other things – including dreaming about the emotional rewards of the venture. She'll fantasize about what her peers will say and how the press will cover it.

On the other hand, the algorithm will patiently sift through the records of *every* relevant takeover in the last fifty years. It will search for the indicators that predict success or failure. It will search through internal records for strengths and weaknesses of both companies. It will search external databases for economic indicators that could affect the results. It will review the history of every person in both companies and identify personality and cultural factors that could predict success or failure. It will also review *all* the literature and research findings in the area. More importantly, it will do all this work *without bias*. It doesn't care if the takeover goes ahead or not. It has no emotional stake in the result and it never gets bored by the details. It makes no assumptions until the data has been reviewed and analysed.

It then produces a decision that's as close to optimal as is possible in an uncertain world. Once you consider what's involved, you have to ask: "How could a human possibly compete with that?"

Next Steps

At the present rate of progress, it's inevitable that something like this will happen – the only question is "When?" It would be very surprising if a significant number of business decisions weren't being made by AI by the year 2028. Perhaps *all* will be made that way by 2038? In other words, the human CEO may have less than 20 years left!

Can it be stopped? In the long term, probably not. Maybe some unforeseen event will slow AI development, but it's hard to see how it could be prevented. For example, even if all Western countries banned it, research would continue in other, non-aligned, states. Even if *every* country banned it, individual research projects would probably continue in secret.

So what can *you* do? The short-term answer is to work with AI systems to maximize your organization's results. Harness them to work on data collection and analysis. Let them make recommendations while your team takes the final decision. That will give you a chance to understand how they "think". To identify their strengths and weaknesses. You may even be able to identify some areas where a human touch is still required!

FURTHER READING

Harari, Yuval Noah, *Homo Deus: A Brief History of Tomorrow*, Vintage.

22 CORE TOOLS

In this chapter, we'll look at tools that'll help you do your job more *effectively* and *efficiently*. They'll help you overcome many of the problems we've discussed. We're going to start with the most important feedback tool – *the decision log*. This is a book that *you* maintain. It contains the details of *all* significant decisions made in the organization. Let's have a look at the format of a typical entry:

Decision no.: 334

Refers to: None

Date: 15/10/18

Name: Boost sales of the Luni-Copter product line.

Trigger: Sales have been flat for the past 23 weeks.

Reason for decision: We need to increase sales because older products are being phased out in 2020.

Decision: Create a new TV advertising campaign for the line. Support it with social media advertising.

Decision details: (Attached)

Estimated cost of implementation: €1.3 million

Top 3 alternatives: (1) Develop new product (2) Billboard advertising (3) Store promotions

Total options investigated: 11

Decision method: Decision matrix

Attachments: Full decision matrix & support materials

Decision group: Joan Lockland, Matt Petri, Tom Cronin, Mary Vy

Expected outcome: Sales increase from €265K to €380K/month

Date when outcome will be achieved: 20/6/2019

Confidence level: 90%

Assumptions: (1) Market for Luni-Copters will remain static (85%). (2) We will capture an increased market share (75%).

Trip-wires: 15/1/19 – (10% increase), 1/3/19 – (20% increase)

MONTHLY REVIEW
> 1. 12/10/18 On schedule.

FINAL REVIEW
Actual results:
Reason for results:
Was the outcome achieved (Y/N)?
Was it achieved on time (Y/N)?
Was it achieved within budget (Y/N)?

Log Overview

You should use the log to record *all* significant decisions. The example above illustrates the launch of a new project. If you decide to shut down the project or modify its goals, its timescale, or its allocated resources, you should record that as a separate decision. It will have a separate decision number and the "Refers to" will reference *this* decision (334). You should also fill out the final review section of this decision to explain *why* the change was necessary.

The most important sections are the "Decision", "Expected outcome", "Date when outcome will be achieved" and "Confidence level". The outcome must be precise. It's not sufficient to say that "sales will increase" or "productivity will improve". You should specify the current and end condition (from/to) in numeric terms. You might argue that isn't possible for everything. What about a project to improve morale, for example? However, if you can't quantify the situation, how can you be sure you've achieved *anything*? How can you be sure there was a problem in the first place? At the end of the project, you'll complete the "Final review" section. The answers to the last three questions will determine whether the project was a success or not.

Why?

Why is the log necessary? Because it provides *feedback* for your decision-making process. We've seen that human memory is malleable. You forget or downplay unpleasant events and remember, at best, a distorted version of reality. That means you don't really learn from your mistakes. The log gives you an *objective* record of your original intentions and the actual outcomes. You may still try to downplay the results ("it would have worked, if only everyone had done their job") but you can't *forget* about the failures.

If you find yourself continually downplaying failures, you'll realise that something isn't right. That should prompt you to change the way you make decisions because you can't keep blaming external factors (other people, customers, the economy, the weather). *You should have taken those factors into account when making your decisions* and if you didn't, *you're* responsible – it's your mistake.

Decision Log Implementation

Start implementing the decision log *today*. Get a hardback notebook and use the headings above as a starting point. You can add or remove sections as you gain experience. If you're not using the decision matrix yet, record your current decision-making method (e.g. personal decision, group discussion and personal decision, senior managers' decision, etc.). Eventually, the log will confirm the need for a more formal approach.

Log Propagation

The next step will be to get your *managers* to keep a decision log. They must record all the decisions they've facilitated with their staff. You'll then be able to review the results and ensure they also have an effective decision-making system in place. However, it's important to work the system *yourself* before asking other people to use it. You need to become familiar with the subtleties and the benefits, so you can justify the effort involved.

You can compare all this effort to putting petrol in a car. You may not like doing it, but if you don't, there *will* be consequences and it'll take a lot more time and effort to deal with them!

Time Management

We've seen that your time is expensive, so you need to use it effectively and efficiently. You need to complete your own tasks and make sure that the agreed strategies and goals are being worked on. You also need to be sure that managers and employees are making the best use of their time. There's a correlation between *your* time management and the total productive time of everyone else in the organization. This is best illustrated with a story:

Dan Jones is the CEO of a pharmaceutical company and *doesn't* have an effective time management system. He often forgets about scheduled meetings, sometimes he arrives late or doesn't turn up at all. He's also in the habit of calling emergency meetings, giving his direct reports only a few hours' notice. No matter what the original subject of the meeting, he confuses things by questioning attendees about operational issues and is visibly upset if the information isn't available.

Dan isn't aware of the chaos he's causing in his organization. When he misses a meeting, his managers are left sitting with nothing to do. They may have spent hours preparing for the meeting and, in the process, distracting team leaders from *their* work to get status updates – all wasted time.

It's even worse when he calls an emergency meeting. Now managers are scrambling to get information at the last minute. They must pull team leaders away and cancel scheduled meetings (some with people outside the company). The irony is that *those* delays will cause more emergency meetings in the coming weeks!

Dan is not willing to let his managers handle problems in their own areas – he *must* be involved. He often becomes immersed in technical issues, where he *has* considerable knowledge.

Dan's case illustrates how a CEO's poor time management and lack of focus can impact *everyone* in the organization. Hundreds of hours are lost because he doesn't respect his own time or anyone else's. His lack of discipline impacts the organization's results. There are two problems: first, Dan is disorganized and wastes time – so he's not *efficient*. Second, he's working on things that should be done by others, and his own work isn't being done – so he's not *effective*. It's possible that his technical knowledge *will* help fix some problems faster, but he's not spending time on his own job, and that's more important. He's also preventing managers and team leaders from developing their own skills.

Guidelines

The CEO's first responsibility is *to manage her own time*. Her next step is to make sure that her managers are managing *their* time properly. The amount of lost time in most organizations due to organizational issues is shocking. Part of the problem is there are no guidelines for managers and team leaders. Many CEOs feel that time management should be left up to the individual, but the cost of that policy is exorbitant. *Management time is a valuable resource that must be administered properly.*

The guidelines in this chapter apply to *everyone*. Some CEOs think they can get their staff to use time management tools but don't bother themselves. That never works out well.

Implementing Goals

Some management teams spend weeks setting new goals. Eventually, the goals are circulated to managers and team leaders. Then everyone ignores them until the first review three months later. That review shows things aren't on track but there's a lot on everyone's plate – many pressing issues – so target dates are adjusted. The same thing happens again and again and eventually, the goals fade away, to be resurrected the following year, perhaps in a slightly revised form. If a goal is important, it must be *operationalized*. That means it must be broken down into action items that can be completed within a few days. Those items must be scheduled and there must be an immediate follow-up on any critical items.

Consistent Goals, Projects and Workflows

Circumstances change. The goals you set nine months ago may no longer be relevant, or the plans you put in place to achieve the goals are no longer valid. Changes can happen rapidly and it's important to recognize when they happen and to respond quickly. For example, employees may be wasting time improving an old process, even though it'll disappear when a goal is achieved. That's why goals need to be synchronised with existing projects and workflows, and any overlaps or contradictions identified and eliminated. That will allow all resources to be put to better use.

Management Tools

The next tool has been designed to help you manage your time and achieve your goals. It's a *time management system* with six elements:

1. A place to plan your time
2. A place to track what actually happened
3. A place for action items
4. A place to track progress on delegated items
5. A place to record your thoughts and ideas
6. A place to highlight your goal of the day

Let's have a look at a typical system in practice:

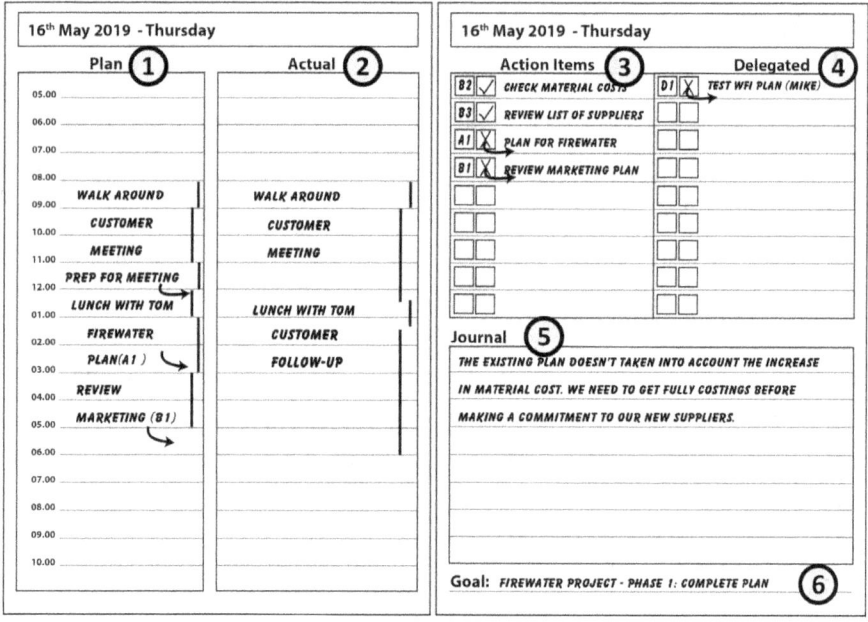

Section 1

This is where you plan your day. If you want to schedule a meeting on the 23rd June, you go to that page, check that you don't already have something scheduled and then record the meeting in that section. If you have a meeting coming up, this is where it'll appear. *If you want to work on something by yourself, you'll also block off time for it here.* It's best to plan your entire day on the previous evening.

Section 2

This is what *actually* happened during the day. People often spend lots of time in their comfort zone (e.g. troubleshooting) and little time in areas where their knowledge or experience is low (e.g. strategic development). CEOs are particularly at risk because there's nobody to look over their shoulder. This section will make sure you're not overlooking any important areas. You may not like what you find, but it *should* prompt improvement.

In the example, you can see that the 9.00am customer meeting ran on for longer than planned. You scheduled a two-hour meeting, but it lasted for three and a half. That meant that the preparation work at 12.00pm had to be postponed and lunch was pushed out by half an hour. You ignored the schedule and held a customer follow-up meeting after lunch, and that meant the items planned for 1.00pm and 3.00pm also had to be postponed (an arrow after a schedule item indicates that it's been rescheduled for later). If you're frequently having to change your schedule like that you should be asking yourself what's causing it – and take steps to prevent it.

Section 3

This is where you record your action items – the tasks *you* must complete. If you try to do too much, you'll burn yourself out and become a bottleneck in your organisation, so you must some reject some tasks and prioritise others and build them into your schedule. To do that, you must distinguish between *strategic*, *urgent* and *routine* tasks and concentrate on those that make the best use of your time.

There are two boxes to the left of each action item. Use the first box to identify the type of item, using the following labelling system:

- "A" indicates strategic tasks
- "B" indicates urgent operational tasks
- "C" indicates routine tasks
- "D" indicates delegated tasks

When you're planning your day, work through your list of action items and label each task using the list above. Cross out any "C" task on your list (they should be completed by someone else) and then sequence all those with the same letter by adding numbers. For example, you might have three strategic tasks. Make those A1, A2 and A3. *You must then allocate time for each item in Section 1.* At the end of each day, you'll tick the second box when an item is complete or put a cross in it if you've failed to complete it. You can reschedule an item by writing it on a future page and drawing an arrow in its box.

Section 4

This is where you'll track your delegated items. You need to check that each item is on schedule. These items will usually link to your strategic goals.

Section 5

Think of this section as The Captain's Log. It's your record of what's happening in the organisation. It captures your thoughts about current events and your plans for the future. It frees your mind from clutter because you can capture your best ideas and, later, go back to the journal if you forget anything. You can't trust your own memory but if you've made a note of your key experiences, you can always set yourself straight. That can save a lot of worry and misunderstanding. It can also give you a very clear idea of what's working – and what isn't.

Section 6

This is where you'll record the strategic "goal of the day". As you write each goal down, you should ask yourself three questions:

- Does this goal still make sense?
- Am I making progress to achieving it?
- What can I do today to move closer to achieving it?

Plan Ahead

Once you have a practical time management system in place, you can leverage it to become more effective. At the start of each week, you should review your strategic goals and identify any actions required to progress them. Then you can schedule those action items and reserve time for them in the "plan" section. That way, you'll be able to make continual progress to achieve them.

Review

When you start to track your time, you'll get valuable feedback and that will help you improve – but only if you're willing to use the information. For example, what percentage of action items did you complete last week? Were there any common trends? Did you complete the operational items and avoid the strategic ones (or vice-versa). Should you cut back on the number of items and delegate more? Did you work on the things you planned, or did you get distracted? Once you start to ask these questions, you'll find areas for improvement.

Your Personality

Some CEOs and managers have problems using a time management system. They argue it's too mechanical, too cumbersome, too restrictive. The same objections could be made about driving a car (why do we need all those rules?) but most managers seem to manage that OK. *An effective time management system is essential.* It will ensure that critical items are addressed, goals are achieved and strategies work. It's an indispensable tool to engage your System 2, and every CEO and manager *must* use one if they hope to succeed.

Email

Humans are easily distracted and that's why you'll need to address one of the major culprits – email. A study by an international IT company found that the average employee spends 40% of their working time dealing with internal emails *that add no value to the business.* That's shocking enough, but there's also an additional psychological cost. What happens when we're monitoring email while trying to create a presentation or write a report? The shift of attention from one task to another disrupts our concentration and saps our focus. A University of London study found that emailing and texting reduces mental capability by an average of 10 IQ points! Studies also show that the effort involved in managing emails leads to increased stress levels. And, finally, to add insult to injury, email is also a major cause of miscommunication. We assume, when we've sent an email, that the recipient has read and understood the message. That's often a mistake. First, it may be lost in an overflowing inbox. Second, even if it *has* been read, it may have been misunderstood because it doesn't contain the cues, like facial expression, tone of voice and gesture, that help us interpret interpersonal messages. Email impairment is such a serious problem that you need policies and systems to reduce its negative impact. Here are some suggestions:

First, consider your own situation. Perhaps you shouldn't have your own email address? Instead, all messages meant for you will be sent to your personal assistant. He'll review them and send on only the relevant ones. However, *he'll only do this once a day at 1.30pm.* You can schedule time from (say) 1.30 to 2.30 to deal with emails. You won't get new emails at any other time of the day, so you can concentrate on your work. What about everyone else? You can get your IT department to hold their emails, and deliver them at 1.30pm. That means everyone can schedule the same time for reading and answering emails and they can concentrate on work for the rest of the day.

You could also limit the number of emails that anyone can send – and you can limit the number of *people* they can send them to. For example, employees might be limited to 300 emails and 900 recipients (an average of three per email) per month. That should reduce the number of pointless emails being sent. You can also draw up policies to reduce the number further. For example, you could ban the use of CCs and BCCs and insist that emails are only sent to those who are *actively* working on the subject of the email. Of course, this approach may not work for everyone. People who work directly with customers and suppliers may need instant access to email. However, the same rules could still apply to their internal email. Right now, it's *certain* that email is sapping your personal efficiency, perhaps by as much as 40%, and it's doing the same to your entire organization. Can you afford to ignore it?

Meetings

Meetings are expensive. Don't hold them unless they're *absolutely* necessary because a large percentage are just a waste of time. If you need convincing, sit in on some of your manager's meetings and ask if they *really* justify the cost. Remember, it's *your* responsibility to ensure that *every* meeting in the organization is efficient and effective. Here are some rules that can make a positive difference:

- Each meeting must be *essential*
- Keep it as short as possible – never longer than an hour
- Keep it small – include only essential people
- Book the room and make sure it's suitable
- Give at least two days' notice for each meeting
- Circulate the agenda the day before
- Circulate objectives – what decisions will be made?
- Ask everyone to prepare beforehand
- Ensure that everyone has accepted the meeting
- Don't hold it if an essential person is missing – reschedule
- Start the meeting on time – flag latecomers
- All phones, tablets and laptops must be turned off
- One person must take notes at the meeting
- One or more decisions *must* be made
- End it on time – use a timer to give a five-minute warning
- Circulate the minutes on the same day
- Avoid scheduling "emergency" meetings (three or more in six months is out of control!)

There's an art to ensuring that meetings work well. The following guidelines will help:

- Everyone gets a chance to speak
- Ideas should be discussed and probed
  – but not attacked
- When anyone makes a factual statement
  – look for evidence
- Don't allow anyone to dominate the meeting

Productivity

In this chapter, we looked at some tools you can use to improve productivity in your organization. It needs to be stressed again that any improvement must start with *you*. You know that most people will resist change (remember the Kubler-Ross model and the 1-8-1 tool?). If *you're* not willing to change, no one else will do so either.

23 EXECUTION

This is the most dangerous chapter. When you finish it, you might just close the book and forget about everything you've learned. Yes, you *want* to become a World Class Leader and you'd *like* to implement those techniques, tactics and strategies, but the whirlpool of daily life is waiting to suck you in. If that happens, your System 1 will make sure you'll continue as before, stuck in the same rut. You really need to act *right now*.

Professor Peter Gollwitzer of New York University has discovered a powerful way to break daily patterns and make changes – it's a technique called *implementation intentions*. Here's how it works: decide what you want to do and then specify exactly *when*, *where* and *how* you'll act. For example, "Each weekday at 8.00am, I'll sit down at my desk and plan my entire day." Or: "Each weekday, just before going home at 6.00pm, I'll sit at my desk, review what happened and identify improvements that can be made." This works because it taps into a subconscious system in our brains that triggers direct action based on environmental cues.

Stop for a few minutes and decide what changes you want to implement. Make a list – and start to implement that list *today*. We'll look at another technique to help you modify your behaviour later in the chapter.

Uncomfortable

Did you find reading this book a little uncomfortable? That's good, because the questions were designed to *force* you to use your System 2. That took effort on your part, but it also *deepened* your understanding and made it more likely you'll change your behaviour. You can use the same technique to get your staff to use their System 2. Of course, you already ask them questions, but if you want people to "think", you need to ask different ones. For example, if you ask the production manager,

"Is everything OK?" you're likely to get a System 1 answer. On the other hand, if you ask her, "What's the percentage increase in throughput relative to this time last year?" then she must *think* about her answer.

Myth Debunking

So, have you been *thinking* about your organization? After all, we've seen that many widely held management beliefs are myths:

- *Interviews* are the best way to select people
- People will learn on *training courses*
- *Experience* is a good indicator of knowledge
- *Gut feel* is the best way to make decisions

Why do CEOs continue to believe and perpetuate these myths when it's easy to prove they're not true? The reason is they don't stop and think for themselves. They do what others do and say what others say. They're *followers*. And, by definition, if someone is a follower, *he's not a leader* – and certainly not *a World Class Leader.*

Strange People

As a CEO, you must work with and influence other people, but there *is* a major problem. A lot of them are *very* strange. You see this on the road when motorists pass you at crazy speeds. They seem to have no regard for road conditions.

To make things worse, you also meet people who crawl along, slowing everyone down. You have to question if these people *ever* use their System 2. And it's not just motorists. What about people in the office? Someone is always complaining that it's too hot or too cold, even when it's perfect. And politics? The people on the right are completely crazy – their policies have become blatantly selfish. But the people on the left are just as bad. They don't seem to understand the political realities any more. How can you be expected to work with people who think like that?

If you agree with any of those sentiments, you're not alone and there's a very good reason for it. You might remember that each of us gathers information about the world using *sensors*. We have eyes that detect electromagnetic radiation with wavelengths from 390 to 700 nanometres. We have ears that detect audio vibrations from 20 hertz to 20 kilohertz. We have a nose that detects odour molecules and we have sensory receptors over our body that can detect temperature, pressure and pain.

Our brain never interacts with the external world directly – it only does so through those sensors. That means when I "see" the colour red, I might have a different experience in my brain from you. We can both agree that an object is red but that means different things to both of us. The same is true for *all* our sensory experiences.

Now, close your eyes for a few seconds. Your knowledge of the world doesn't disappear. You know where your car is parked and how your kitchen looks even though you can't see either one. You have a model of the world in your head – but that model isn't perfect. You probably don't remember how many tiles cover the kitchen floor or the pattern on the car seats. *Your model of an object is different from other people's model of the same thing.* You remember different things about it. And it's not just physical objects; we also create models of non-physical things – like behaviour.

How do you decide the right speed to drive at? Do you construct an elaborate equation using the ambient temperature, the road surface, the traffic density, depth of tyre thread, the condition of your shock absorbers and *then* calculate the right speed? Of course not – your System 1 makes the decision based on an *internal* model of what's right. It's the right speed because *you've* chosen it and anyone driving faster or slower is wrong. That's called *naïve realism.*

You drive at the perfect speed, you adjust the office temperature until it's perfect and your political stance is perfectly balanced between right and left. You're right about *everything*. Of course, it's just an illusion but we all suffer from it, and it's a major reason why we find it hard to agree on things. If your model is different from mine – and you believe you're *always* right and so do I – then we can never agree. As CEO, you need to be conscious of your tendency to depend on *naïve realism* when you're working with people. It seems perfectly obvious that *your* views are correct, but others are using different models. And it's possible – occasionally – that one of them might be closer to reality.

Denial

We've seen that people suffer from biases that cause many of their decisions to be flawed. One of the most common is the tendency to continue a practice even though it's been shown to have negative consequences or is counterproductive. For example, some people refuse to exercise, even though they know it would improve their health (an act of omission). Others continue to smoke even though they know it's damaging their body (an act of commission).

These are acts of *denial*. These individuals refuse to acknowledge the consequences of their actions (or inaction) and, instead, search for any hint that the situation isn't as bad as it seems.

World Class Leaders can't allow themselves the luxury of these delusions. *But do you?* You've learned that many of your systems are flawed. It's likely that your recruitment system isn't selecting the best people and your training isn't giving employees the knowledge, skills and motivation they need. Your email system is killing productivity and your meetings are wasting resources. Most significantly, your decision-making process isn't producing the best decisions. What are you going to do about it? Remember, doing nothing *is* a decision – but not a very good one.

Write down your answer before reading on (Q23-1).

I've already pointed out that if you finish the book and put it aside – even with the intention of getting back to it in the future – you've made a decision *to do nothing*. But perhaps you're still not convinced by some of the arguments in this book? That's your System 1 asking your System 2 to justify the status quo. Still… let's say that *is* how you feel. If you're not convinced, then you have a responsibility to dig further and *find the truth*. If this book is right, you have an enormous opportunity to improve. Are you going to let System 2's laziness stop you?

The right course of action is to act *now*. Make a start. Block off time in your planner to act today, tomorrow and for the rest of the week.

If you're not sure about something, test it in your own organization. For example, if you can't accept that your selection process is flawed, analyse the performance of the people hired in the last year.

Leadership Beliefs

We've talked a lot about leadership. Now let's see what you *really* believe. In Chapter 3 (Q3-2) you were asked:

- Do you regard yourself as a strong leader?
- Do you expect your managers to follow your lead?
- Do you expect all your staff to do as they're asked?

Look at what you wrote down. Did you agree that you *are* a strong leader, and did you answer "yes" to questions 2 and 3? If so, you probably see yourself as *a benevolent dictator*. You make the decisions – everyone else does what they're told. But how did you answer the questions in Chapter 15 (Q15-1)?

- Are you good at delegating?
- Should managers usually make their own decisions?
- Should workers make decisions for themselves?

Did you agree that you *are* good at delegating? And did you also answer "yes" to questions 2 and 3? If so, there's a conflict between the two sets of answers – *how can you be a strong leader if you let people make their own decisions?* And then you were asked three more questions in Chapter 10 (Q10-7):

- Do you *actively* listen to people before coming to a decision?
- Do you *actively* look for alternative points of view?
- Are you willing to change your mind if someone makes a good point?

If you said "yes" to these three questions, it paints a picture of someone who requires a *lot* of input before they can make up their mind. Someone who's *not* willing to make a decision and stick to it. It seems like the *opposite* of your initial answers. Are you *confused* about leadership?

The Leadership Myth

So, what *is* a "strong leader"? To many people, it's someone who makes *all* the important decisions. A person who doesn't worry about what others think or the possible negative consequences of his decisions. Unfortunately, that definition best describes individuals like Adolf Hitler, Joseph Stalin, Mao Zedong, Benito Mussolini, Saddam Hussein, Muammar Gaddafi, Idi Amin and many more – and that's no coincidence. Strong leaders end up believing that they know better than everyone else. They gloss over their lack of knowledge and ignore their own mistakes.

The historian Lloyd Clark paints a chilling picture of the disarray that these people create:

> *"The generals were justifiably concerned at Hitler's lack of consultation [and] his oversimplification of a whole series of massively complicated issues. It seemed that the Fuhrer's mind was already made up and no experts or documents were likely to change it."*

You can see why a strong leader is a disaster just waiting to happen.

The Leadership Secret

But perhaps you mean something else when you talk about a *strong leader*? This book uses an alternative term: the *World Class Leader*. A World Class Leader must do many things:

- Take responsibility
- Communicate
- Motivate
- Mentor
- Gain people's trust
- Create a strategy and a vision
- Manage the business

But while these are all necessary, they're not sufficient. Underlying all of them is something far more important – and it's the secret of leadership. A World Class Leader will do the one thing that others are reluctant to do. *She'll use her System 2 more than anyone else in the organization.* Why? Because she *knows* that she suffers from all the human biases we've talked about. She knows that System 2 is the key to achieving her goals. Many CEOs *aren't* willing to do that. They accept waste, functional stupidity and bogus strategies. They maintain the status quo rather than creating progress. They're comfortable making decisions using a System 1 programmed by fairy tales, random encounters and other people's influence.

Of course, it's not easy. Work out 5 + 6 and it's effortless – that's System 1. But what happens when you try 68 x 47? Yes, System 2 is difficult and uncomfortable. It's almost impossible to use it all the time, but there *is* a solution. You can *reprogram* your System 1 to act in a way that's dictated by your System 2. We already touched on this when we created a list of responses for potentially emotional events, like someone missing their sales targets. The diagram below shows how to reprogram yourself to become a leader, rather than just following other people's agendas:

First, you identify the new *behaviour* you want to install using de-cision-making tools. You also identify a *trigger* for the behaviour (time and location) and determine exactly *how* it's going to be carried out (the *procedure*). Then you repeat the behaviour *consciously* when the trigger occurs. Each time you succeed, you give yourself a small reward. Eventually, after perhaps twenty or thirty repetitions, it becomes an automatic (System 1) response.

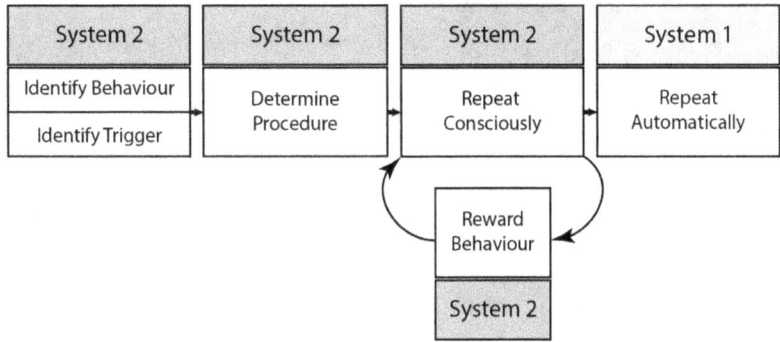

You may be reluctant to include that reward – after all, you're *not* a child. In fact, your System 1 *is* child-like and the reward could make all the difference between success and failure.

Here are a few recommendations for that reward:

- It should be desirable
- It should be relatively small
- It shouldn't be normally allowed.

For example, you might allow yourself a single Ferrero Rocher chocolate each time you succeed – but don't indulge at any other time. You can use that approach to embed the *decision log*, the *time management system* and the *decision matrix*. When you use this reprogramming technique, you're embedding System 2 behaviours into your System 1. You're deliberately deciding what you're going to do in the future.

As a *leader*, you *must* be the model for the use of System 2 in your organisation; otherwise, everyone else will avoid using it. Even experts avoid it by specialising in narrow niches. But you can lead by example. You can change the way decisions are made – not just at management level, but at *every* level.

Presentation

By now you'll have accepted that your knowledge is limited, your decisions are imperfect, and your thinking is distorted by *naïve realism* and System 1 thinking (unless you're still in denial?). Should you allow this realization to affect how you present yourself? Should you appear doubtful and hesitate when making public statements? Of course not – that would be a huge mistake.

When you speak, you must project *confidence*. You must reassure people that the organization is going in the right direction. You must *inspire* others. However, many good CEOs fail to do that. They *think* they're being positive and persuasive but are actually spreading unease and doubt. So, what can *you* do to get the right message across? Essentially, *you must meet people's perceptions of a strong leader!* Yes, unfortunately, you must feed the myth.

Let's start with your *body language*. Stand with your spine straight, your shoulders back and your head high. Men should position their feet under the shoulder joints while women should position them under the hip joints. Your weight must be balanced evenly on both feet. Your head must be level and your nose should point directly at the listener. You should gesture with your hands, but never hold them in front of your face.

What about your speech? Talk clearly and reasonably slowly. Enunciate each word carefully. Never use filler words (like "ah", "umm", "basically", or "you know"). Vary the pitch, pace, inflection and emphasis to match your message. Your speech and body language will make an enormous difference to your credibility. If you have doubts, look for the excellent clip of Cara Hale Alter on YouTube (search for: "Cara Alter, Credibility Code, Talks on Google").

Some CEOs believe they can stand up and talk without any preparation. That can lead to embarrassing moments. You *must* be prepared and that means starting with what you want to achieve – what is your objective? You should have *one* clear goal in mind. Next, you must decide *how* to achieve it. You should be able to answer the following questions:

- Who'll be in the audience?
- Who are the decision makers?
- What are they interested in?
- What's in it for them?

You can then start to develop your script – write down all the points you can think of, and then start pruning. There are a few guidelines to get the best results:

- The talk should last *less than* 20 minutes
- Limit yourself to *three* key points
- Use short *stories* to illustrate your points
- Using PowerPoint? *Use less than* 10 words per slide
- Use photographs and *simple* diagrams instead of words
- Be prepared for questions

The most important rule is to *practise* your talk. Practise in front of a video camera and then play it back. Practise it again, to make sure you're standing properly. Practise it again, to make sure you're speaking properly. Practise it again, to make sure the timing is right. Practise it again, to make sure you're covering the key points and getting your message across. The more important the talk, the more you must practise. If it's a *very* important presentation, you should also practise in front of a small audience and get their feedback. How you present yourself, what you say, and how you say it, will make a big difference to your reputation as a leader.

Confidence

It's unfortunate that many of the most confident and forceful people *know very little*. They believe things are simple when they're not – they don't appreciate the complexities of the situation. Often, they don't even realise they're making mistakes. But, occasionally, some of these people can be successful, and part of the reason is that *their confidence can inspire others*.

While confidence can be powerful, it can't overcome *every* obstacle. Let's say you're confident that we're going to have a hot summer, so you invest in a huge shipment of beachwear. Will your confidence bring success? Of course not, because your belief has no effect on the weather and there's no reliable way of predicting it months ahead.

On the other hand, if you're confident that you can increase production by 10%, that enthusiasm *could* motivate people to put in extra effort. It could break down barriers and trigger novel solutions. It could be the catalyst to achieving those results. So confidence is important, but *informed* confidence is always better than *blind* confidence. If you use the forecasting and decision-making techniques from this book, you *can* be confident that your decisions have the best chance of success.

Your Purpose

We're almost at the end, so it's worth considering: *why* you do what you do? Obviously, you want to make money so that you and your family can live in comfort and enjoy a few luxuries. But is that all? Do you have a greater purpose? Remember – you're in a position of power. CEOs *control* most of the world's wealth, so you can do much good. There are many injustices you can help redress. You *know* about the thousands of people dying from starvation around the world. You know about infants suffering from horrible diseases that *could* be cured.

What can *you* do – not just as an individual, but *as a CEO* – to make the world a better place?

Many executives just turn a blind eye, even when they could influence the situation. For example, are you comfortable ignoring what happens in many *factory farms?* Are you aware that baby calves are dragged from their mothers as soon as they're born and stuck into pens so small that they can't even turn around. They're confined in misery, never getting the chance to walk, run or play, never getting a chance to live a natural life. The same happens to pigs, chickens and other animals, who are stuffed into wire cages without room to move and stacked in filthy, windowless sheds. Those animals suffer unimaginable physical and emotional agony every minute of their short lives. Why? Just to improve the bottom line. You could help change these things but are you willing to take the first step? If not, what's the point in being a leader?

The Objective

This book was written to help you become a *World Class Leader* and I hope it's already begun working. Have you implemented the decision log and the time management system? Have you started tackling areas that you previously overlooked (like recruitment and training systems)? You should have many new ideas, and the techniques to implement them. The important thing is that you *do* something different – and that you start right away.

> **We're coming to the end, but don't close the book – go back and read it again. You'll learn much more the second time.**

For example, did you notice that critical piece of advice about mindset in Chapter 15? We were discussing the recruitment process, but it also applies to your existing staff – and yourself. You're more likely to pick it up on your second reading.

The first reading was primarily designed to shock your system. It should have jolted you out of your rut (the status quo) and explained why decisions you *thought* you were making have been programmed by others. Up to now, you've been a follower - *especially when you felt most like a leader.* Now you know how to make good decisions and the next time you read the book, you'll identify even *more* ideas and concepts, many of which could have a cathartic effect on your organization. You *can* become *a World Class Leader*, and I hope you will.

If you have any questions, you can go to our business website at http://www.wclass.com or my personal one at http://www.pat-hough.com. Please let me know how you get on by leaving a message. Thank you for reading to the very end.

May the odds be in your favour – but don't depend on it!

FURTHER READING

Alter, Cara Hale, *The Credibility Code: How to Project Confidence and Competence When It Matters Most*, Meritus Books.
Gilovich, Thomas & Ross, Lee, *The Wisest One in The Room: Think Clearly. Make Better Decisions. Influence People*, Oneworld Books.
Barg, John, *Before You Know It: The Unconscious Reasons We Do What We Do*, Windmill Books.
Clark, Lloyd, *Kursk: The Greatest Battle: Eastern Front 1943*, Headline Review.

QUESTIONS AND ANSWERS

If you'd like to get feedback and scoring on your answers to the questions posed in this book, please go to http://pat-hough.com.